HEALTH, DISEASE, AND CAUSAL EXPLANATIONS IN MEDICINE

PHILOSOPHY AND MEDICINE

Editors:

H. TRISTRAM ENGELHARDT, JR.

Center for Ethics, Medicine, and Public Issues,
Baylor College of Medicine, Houston, Texas, U.S.A.

STUART F. SPICKER

University of Connecticut Health Center, Farmington, Connecticut, and
National Science Foundation, Washington, D.C., U.S.A.

VOLUME 16

HEALTH, DISEASE, AND CAUSAL EXPLANATIONS IN MEDICINE

Edited by

LENNART NORDENFELT

Linköping University, Linköping, Sweden

B. INGEMAR B. LINDAHL

*Huddinge University Hospital, Huddinge, and
University of Stockholm, Sweden*

D. REIDEL PUBLISHING COMPANY

A MEMBER OF THE KLUWER ✦ ACADEMIC PUBLISHERS GROUP

DORDRECHT / BOSTON / LANCASTER

Library of Congress Cataloging in Publication Data
Main entry under title:

Health, disease, and causal explanations in medicine.

(Philosophy and medicine; v. 16)
Papers presented at the First Nordic Symposium on the Philosophy
of Medicine, held May 13–15, 1982, in Stockholm.
Includes bibliographies and index.
1. Medicine–Philosophy–Congresses. I. Nordenfelt, Lennart,
1945–   . II. Lindahl, B. Ingemar B. (Börje Ingemar Bertil),
1949–   . III. Nordic Symposium on the Philosophy of Medicine
(1st : 1982 : Stockholm, Sweden) IV. Series. [DNLM: 1. Health–
Congresses. 2. Disease–Congresses. 3. Philosophy, Medical–
Congresses. W3 PH609 v. 16 / W 61 H43415 1982]
R723.H395   1984   610'.1   84–4767
ISBN 90–277–1660–9

Published by D. Reidel Publishing Company,
P.O. Box 17, 3300 AA Dordrecht, Holland.

Sold and distributed in the U.S.A. and Canada
by Kluwer Academic Publishers,
190 Old Derby Street, Hingham, MA 02043, U.S.A.

In all other countries, sold and distributed
by Kluwer Academic Publishers Group,
P.O. Box 322, 3300 AH Dordrecht, Holland.

Printed in The Netherlands.

To the Memory of
Ingemar Hedenius and Clarence Blomquist
Two Pioneers in Swedish Philosophy of Medicine

TABLE OF CONTENTS

EDITORIAL PREFACE   xi

LENNART NORDENFELT / Introduction   xiii

SECTION I / ON THE CONCEPTS OF HEALTH AND DISEASE

INGMAR PÖRN / An Equilibrium Model of Health   3

LENNART NORDENFELT / Comments on Pörn's 'An Equilibrium Model of Health'   11

LENNART NORDENFELT / On the Circle of Health   15

STUART F. SPICKER / Comments on Nordenfelt's 'On the Circle of Health'   25

H. TRISTRAM ENGELHARDT, JR. / Clinical Problems and the Concept of Disease   27

KAZEM SADEGH-ZADEH / Comments on Engelhardt's 'Clinical Problems and the Concept of Disease'   43

RALPH GRÄSBECK / Health and Disease from the Point of View of the Clinical Laboratory   47

SECTION II / ON DEFINITION AND CLASSIFICATION IN MEDICINE

UFFE JUUL JENSEN / A Critique of Essentialism in Medicine   63

HENRIK R. WULFF / Comments on Jensen's 'A Critique of Essentialism in Medicine'   75

HELGE MALMGREN / Psychiatric Classification: The Status of So-Called "Diagnostic Criteria"   77

STAFFAN NORELL / Comments on Malmgren's 'Psychiatric Classification: The Status of So-Called "Diagnostic Criteria"'   89

SECTION III / ON CAUSAL THINKING IN MEDICINE

A. Causal Analysis

ANDERS AHLBOM / Criteria of Causal Association in Epidemiology   93

GERMUND HESSLOW / Comments on Ahlbom's 'Criteria of Causal Association in Epidemiology'   99

ANNE M. FAGOT / About Causation in Medicine: Some Shortcomings of a Probabilistic Account of Causal Explanations   101

B. Causal Selection

STAFFAN NORELL / Models of Causation in Epidemiology   129

B. INGEMAR B. LINDAHL / On the Selection of Causes of Death: An Analysis of WHO's Rules for Selection of the Underlying Cause of Death   137

ØIVIND LARSEN / Disease from a Historical and Social Point of View: Some Remarks Based on the Needs of Preventive Medicine   153

B. INGEMAR B. LINDAHL / Comments on Larsen's 'Disease from a Historical and Social Point of View'   165

HENRIK R. WULFF / The Causal Basis of the Current Disease Classification   169

H. TRISTRAM ENGELHARDT, JR. / Comments on Wulff's 'The Causal Basis of the Current Disease Classification'   179

GERMUND HESSLOW / What is a Genetic Disease? On the Relative Importance of Causes   183

HENRIK R. WULFF / Comments on Hesslow's 'What is a Genetic Disease?'   195

C. Causal Explanation

KAZEM SADEGH-ZADEH / A Pragmatic Concept of Causal Explanation   201

INGMAR PÖRN / Comments on Sadegh-zadeh's 'A Pragmatic Concept of Causal Explanation'   211

D. *Other Topics on Causality*

ERIK ALLANDER / Holistic Medicine as a Method of Causal Explanation, Treatment, and Prevention in Clinical Work: Obstacle or Opportunity for Development?   215

STUART F. SPICKER AND H. TRISTRAM ENGELHARDT, JR. / Causes, Effects, and Side Effects: Choosing Between the Better and the Best   225

APPENDIX

B. INGEMAR B. LINDAHL / Notes on the Philosophy of Medicine in Scandinavia   237

NOTES ON CONTRIBUTORS   249

INDEX   251

EDITORIAL PREFACE

On May 13–15, 1982, some 50 scientists and scholars – physicians, philosophers and social scientists – convened at Hässelby Castle in Stockholm for the first Nordic Symposium on the Philosophy of Medicine. The topics for the symposium included (1) the concepts of health and disease, (2) classification in medicine, and (3) causality and causal explanations in medicine. The majority of the participants were Scandinavian but the symposium was also able to welcome four distinguished guests from other parts of the world, Professors Stuart F. Spicker and H. Tristram Engelhardt, Jr., U.S.A., Dr Anne M. Fagot, France, and Dr Werner Morbach, West Germany. The latter represented Professor Kazem Sadegh-zadeh, who unfortunately was prevented from attending.

One of the main purposes of this symposium was to bring together people in Scandinavia who at present work within the field of Philosophy of Medicine. This group is still relatively small but is growing rapidly, and the scholarly activity has recently been notable. This fact is clearly demonstrated by the presentation of 'Philosophy of Medicine in Scandinavia' in the *Appendix* of this volume.

We, the editors of the volume, wish to express our gratitude to a number of persons and institutions who made this symposium possible. We would like to thank our financial sponsors, the Medical Research Council of Sweden and the Faculty of Arts at the University of Stockholm. We also wish to thank the Heads of the Department of Philosophy (University of Stockholm), and the Department of Social Medicine (Huddinge University Hospital), Professors Dag Prawitz and Erik Allander, for their advice and for their personal contributions to the symposium. We acknowledge with gratitude the assistance provided by Miss Karin Lundström and Mrs Rose-Marie Hellberg of the Department of Social Medicine during the preparatory work for the symposium.

Special thanks are due to the President of the University of Stockholm, Professor Staffan Helmfrid, who opened the symposium and kindly honoured it by his presence.

We are particularly grateful to the Editors of the book series, *Philosophy and Medicine*, Professors Spicker and Engelhardt, for their offer to include

L. Nordenfelt and B.I.B. Lindahl (eds.), Health, Disease, and Causal Explanations in Medicine, xi–xii.
© 1984 *by D. Reidel Publishing Company.*

the scholarly substance of the symposium in their series. They have also assisted us greatly in the preparation of the volume. So too has Dr. Paul Needham, who helped us by perusing the text, and improving it from a linguistic point of view.

March 8, 1983,   LENNART NORDENFELT
Stockholm, Sweden   B. INGEMAR B. LINDAHL

INTRODUCTION

I. ON THE CONCEPTS OF HEALTH AND DISEASE

The entire medical enterprise — theoretical and clinical research as well as actual medical practice — has human health as its ultimate end. Health, as well as disease and illness, must always be in the focus of medical attention; the basic medical activities such as prevention, diagnosis, therapy, and cure have as their starting points the phenomena of disease and illness and as endpoint the ideal of health. The concepts of health and disease therefore have a self-evident locus in the center of medicine's conceptual network. The other medical notions are parasitic upon health and disease. Satisfactory theories of the former presuppose a clear understanding of the latter.

In spite of their central place, however, and in spite of numerous efforts directed to the clarification of the concepts of health and disease, there is far from universal agreement about their nature. In fact, the controversies are quite profound: they concern such fundamental issues as whether or not 'health' and 'disease' are truly scientific concepts. There are good psychological and historical explanations for this state of affairs. Health and illness are facts of extreme importance to all human beings. We want to understand them as fully as possible and from as many aspects as possible. As a consequence, individuals from different backgrounds and with very different approaches address the phenomena of health and illness. Thus, one encounters anthropological, sociological, psychological and biological theories, as well as combinations of these. The contents of the various theories are quite different and often quite difficult to compare.

Rather recently, analytic philosophers have entered the scene. One of their main tasks has been to disentangle the conceptual conflicts and complications in this arena. This has proved to be quite difficult. The results of the conceptual analyses performed are disparate and, at times, even incompatible. Still, some progress has been achieved. Many of the basic conceptual theoretical frameworks have been explored. Most importantly, there is some basis for making satisfactory comparisons between the various views.

We shall not attempt here to survey the discipline — the Philosophy of Medicine. (For an excellent overview, see [4].) For the purpose of introducing

L. Nordenfelt and B.I.B. Lindahl (eds.), Health, Disease, and Causal Explanations in Medicine, xiii–xxx.

the contributions to this volume we shall present (in a simplified way) two very influential doctrines, which stand in very sharp contrast to each other. The first doctrine rests mainly on a biological platform, the second one mainly on an anthropological foundation. Both doctrines face several problems. They therefore appear to call for a synthesis. A sketch of such a synthesis is, then, provided in the following:

Doctrine (1). Health and disease are attributes of human beings considered in their entirety as well as viewed as parts, for instance, organs and cells. By the phrases "X is in health" and "Y is diseased" we mean, respectively, that the functions of X are normal and that the functions of Y are abnormal. Normality should here be understood in a biostatistical sense: the functions of X are normal, if and only if they are typical for members of the species to which X belongs.

According to this view, 'health' and 'disease' are *descriptive*, scientific terms. They belong to microbiology and physiology, as well as to anthropology and clinical medicine. There are no (and need not be any) values attached to these terms. The fact that most people aim for health and disvalue disease is a fact which is completely independent of the meaning of the terms 'health' and 'disease'.

Doctrine (2). Health and disease are attributes applicable only to human beings viewed as integrated wholes. A cell or an organ cannot properly be assigned health or disease: By the phrases "X is in health" and "Y is diseased" we mean, respectively, that X functions well and Y functions badly. Since we are talking about the functions of persons, 'function' refers to ability to perform actions; the healthy person is able to do what she/he should be able to do; the diseased person is not. According to this view, 'health' and 'disease' are *evaluative* terms. If they should, at all, be called scientific, they belong only to those disciplines which treat human beings as wholes, such as anthropology and clinical medicine.

These sketches are purposively simplified. No philosopher would accept these few lines as adequate descriptions. We might, however, be permitted to say that Christopher Boorse's theory [2] comes quite close to the first position, whereas the theory of Georges Canguilhem [3] more appropriately embodies the second.

How is one to select among these approaches? As we have already indicated, both theories seem to face difficulties. The first view must account for the following facts:

(a) There are deviations from statistically normal functioning, both on the molar and molecular level, which are not properly regarded as diseases or signs of disease.

(b) There are statistically normal phenomena which are regarded as diseases.

The second view must account for the following facts:

(c) Medical science identifies particular, anatomically located diseases. These seem to be attributable to specific organs or systems of organs and not to the person as a whole.

(d) In somatic medicine there is general agreement about what biochemical or physiological states should be viewed as healthy and what viewed as diseased. Does this not indicate that disease language is a purely scientific and non-evaluative language?

For our purposes we shall sketch a synthesis based on the second platform. We offer an account of (c) and (d) within the general framework of a holistic and evaluative view of health and disease.

Doctrine (3). Health is an attribute which is *primarily* applicable to human beings as integrated wholes. In a *derivative* sense, however, one can say that an organ is healthy. (This means that the function of the organ is such that it promotes the health of the person as a whole.)

In order to describe the contradictory of health it may be illuminating to substitute the term 'illness' for 'disease'. When a person is not healthy, then he is ill. The point in doing this is to reserve the term 'disease' for those often anatomically isolated states which tend to cause unhealth or illness.

Objection (c), then, is pertinent in pinpointing the existence of particular diseases. This objection does not, however, imply that a particular disease can be identified independently of its relation to human illness. A disease is not a disease because of any intrinsically biochemical component; nor is it a disease because it is statistically abnormal. X is a disease because X tends to affect human health negatively − it tends to cause illness in its bearer.

This conceptual relation between disease and illness is completely compatible with the fact that diseases, once they have been identified, can be described in purely biochemical or physiological terms. On this understanding, then, a disease may be an object of the specialized medical disciplines. Its analysis may be completely free from evaluative terms.

What then are the criteria for determining a person's health? According to position (2), a person is in health when she/he functions well, i.e., when she/he can perform those actions which she/he should be able to perform. But what should one be able to perform? Here a number of interpretations present themselves: (a) one should be able to perform what the majority of one's cohorts are able to perform; (b) one should be able to perform what one's basic needs require; (c) one should, in normal circumstances, be able to satisfy one's aspirations or ambitions.

If we adopt (c), we are brought to the theory of health proposed by Ingmar Pörn in the first essay in this volume. In 'An Equilibrium Model of Health', Pörn formulates his dominant idea about health in the following way:

Health is the state of a person which obtains exactly when his repertoire [abilities] is adequate relative to his profile of goals. A person who is healthy in this sense carries with him the intrapersonal resources that are sufficient for what his goals require of *him* ([9], p. 5).

Health, for Pörn, thus, is a genuinely relational concept. This relation is one of mutual fitness between repertoire and goal. This idea of an equilibrium is what essentially distinguishes Pörn's view from other representatives of the general position outlined above.

The contradictory of health, in Pörn's system, is illness. This is a state of a person which obtains when his repertoire is insufficient for realizing his profile of goals. The concept of disease (as well as the concepts of *impairment* and *injury*) does not belong to the same dimension as health and illness. The former concept refers to the *biological basis* which is responsible for the repertoire/goal relationship.

A disease, on this account, need not always cause illness. Certain bio-psychological states are called "diseases" because they have a *tendency* to restrict repertoires. They do not have such an effect in every individual case.

But, how, given Pörn's position, can we account for the fact that most societies agree in the identification of diseases on the pathoanatomical level? A reasonable reply is the following: most of the states which we identify as diseased affect such basic components in one's repertoire (such as one's ability to move one's limbs or one's ability to see or hear) that, whatever the needs or aspirations an individual might have, these would be compromised by the disease in question.

Two other contributors to this volume, H. Tristram Engelhardt, Jr. and L. Nordenfelt, plead against the biostatistical conception of health and disease, in favour of a holistic, evaluative view. There is, however, an important difference between the views of Pörn and Nordenfelt and those of Engelhardt. The former have a traditional, philosophical project in mind; they aim at constructing a coherent and plausible theory of health with as simple a set of analytical tools as is possible. In their treatment there are few references to specifically clinical concerns and to the use of medical terms in actual clinical practice.

In contrast to this approach, Engelhardt's 'Clinical Problems and the Concept of Disease' emphasizes the importance of founding the Philosophy of Medicine in clinical practice. The defect of many philosophical analyses of health and disease, he says, has often been their abstractness. Engelhardt therefore suggests a different approach for reconstructing the concept of disease. He proposes a terminological change as one of the ways in which this reorientation could be accomplished. Instead of talking about diseases, defects, or deformities, we should use the general label "clinical problem". This terminological change would then underscore the fact that what is to be identified as the object of medical interest are problems to achieve certain medical goals and not any absolute biological entities.

Clinical medicine is focused on resolving clinical problems: problems with pains, expected human form or grace, and expected abilities to function, insofar as these problems are seen as having a physiological or psychological basis, and as being beyond the direct and immediate volition of the individuals held to "have" the problems ([5], p. 33).

The important relation between conceptual analysis and actual clinical concerns is even further emphasized in 'Health and Disease from the Point of View of the Clinical Laboratory', offered by Ralph Gräsbeck. One of the interesting issues raised by Gräsbeck is the following: Can laboratory medicine provide criteria for health? Are there any health values on the biochemical parameter? Gräsbeck reveals some of the difficulties in providing answers to these questions. First, in order to obtain reference-values characteristic of healthy people, some previous selection of healthy individuals must have been made. But what criteria should be used in this selection? According to Gräsbeck, one has traditionally used the individual's own assessment of his state of health. For many reasons, however, clinical chemists have been discouraged from using only such criteria. A person who considers himself healthy might have hidden disturbances which affect laboratory values.

The tricky question then arises whether one should use the laboratory itself in determining the health status of the reference-individuals. There is a risk of a vicious circle which could be summarized in the following statements: we need a population of healthy reference-individuals in order to determine health reference-values; therefore we need criteria for selecting the right reference-individuals. Some of these criteria involve laboratory tests. But such laboratory tests already presuppose some knowledge of health reference-values, and these were precisely the things to be determined.

How should these difficulties be accounted for within the framework sketched above? On the surface it seems as if clinical chemistry oscillates between two views of health. On the one hand, there is an obvious reference to health on the level of the whole person; and here it is clearly not only a question of statistics. On the other hand, there are cases, according to this line of reasoning, when *molar* health is simply illusory or misleading. Whether a person is healthy or not often depends upon certain data gleaned from the *molecular* level.

Let us briefly suggest how a proponent of position (3) might react to this situation. He might say that there need not be a vicious circle here. There is a rationale in what clinical chemists in fact do. This is so on the understanding that clinical chemists do not, strictly speaking, search for criteria for health; instead, they seek criteria for the *absence* of disease. According to position (3), it is possible for a person to be healthy while at the same time having various diseases. Therefore, we cannot use every healthy person as a reference-individual if we wish to identify criteria for the absence of various diseases.

II. ON DEFINITION AND CLASSIFICATION IN MEDICINE

Our notion of definition stems from Aristotle [1]. His ideas about natural hierarchies, species, and their essences have had, and still have, a profound influence on Western scientific thought. Aristotle maintained that the things to be defined are the *species*. We define the various species by adequately describing their essences. More specifically, a species, (e.g., human) is defined by first giving the *genus proximum* (the most proximate biological order) to which it belongs; in the case of humanity the genus is *homo*. The second requirement consists in giving the *differentia specifica*, i.e., that characteristic which differentiates the species from other species belonging to the same genus; for humanity it is the characteristic *sapientia*, wisdom.

The idea that species have essences and that a definition should contain a complete description of these has, in spite of very forceful criticisms, been very long-lived. Also theoretical medicine, in particular during the era of the great taxonomic systems, has been founded upon the Aristotelian platform. Diseases were looked upon as sharply delineated *species* belonging to natural hierarchies – *genera*. The disease species could, in principle, be defined in terms of its most proximate genus and its differentiating characteristic.

Although such faithful essentialism has now almost completely vanished from the medical scene, there still remain some elements of the Aristotelian theory of definitions. Scholars tend to agree that there are no naturally given disease-species to be defined. The 'diseases' that we talk about are more or less arbitrary inventions. The taxonomic system that currently prevails could have been construed otherwise, with equal or perhaps even better therapeutic consequences.

The fact that diseases are conventionally delimited entities makes it, according to many theorists, even more important to characterize them in explicit terms. This becomes particularly essential for the purpose of meaningful communication. If a Swedish doctor wants to communicate information about the disease *gastritis* to a North American doctor, and no explicit definition of *gastritis* has been agreed upon, it appears that there is great risk that the American would misunderstand this information. He might think that it is applicable to a kind of patient to whom it is not. It seems that although we have abandoned Aristotelian essences there is a great need for explicit definitions of our disease concepts in terms of necessary and sufficient criteria for their application.

In 'A Critique of Essentialism in Medicine', Uffe Juul Jensen also takes issue with this modified requirement concerning explicit definitions. He first points out that current medical textbooks offer no explicit definitions of diseases. What they give could at best be characterized as "ultrashort descriptions," which adequately describe only some very typical exemplars of a disease.

This is, says Jensen, *not* a lamentable state of affairs. The attempt to give something like complete definitions of diseases would be an impossible and misguided enterprise. In arguing for his position, Jensen finds theoretical support in current theory of definition in evolutionary biology. Here an important distinction is made between *biological species* and *taxonomic species*. By a biological species is meant the species looked upon in an historical perspective, as a unit of evolution. A taxonomic species is an idealization of a typical exemplar existing at a particular time. A taxonomic species is a unit of classification and can, in contradistinction to the biological species, be explicitly defined. Using the distinction, one may say that the characterization of diseases used in textbooks delimits diseases as units of classification. But these descriptions do not capture diseases as processes, i.e., diseases as units of evolution.

In commenting on these views, Henrik Wulff expresses some concern with their consequences for communication in medicine. He suggests that the need for explicit definitions – in terms of necessary and sufficient conditions – is greater in medicine than in evolutionary biology. The reason for this is that the inter-observer agreement is not as strong in clinical medicine. Hence, one also needs working definitions – in terms of as complete a list of signs and symptoms as possible – of the various disease entities in their status as units of evolution.

Problems of definition and classification in medicine are the concern also of Helge Malmgren's 'Psychiatric Classification: The Status of So-Called Diagnostic Criteria'. Malmgren, too, questions the requirement of explicit definitions. He discusses the question of whether a set of operational criteria for a diagnosis should be regarded as a definition of the concept(s) used in the diagnosis. His thesis is that operational criteria may be useful in identifying the meaning of a disease name. They need not, however, amount to a definition. On the other hand, a definition of a disease is not a necessary prerequisite for meaningful talk about the disease.

In arguing for his standpoint, Malmgren considers two concrete examples, two rare diseases — *encephalitis lethargica* and *Legionnaire's disease*. These two diseases had, for a long time, the following in common: no specific etiological agent was known; there was no clear-cut clinical picture or pathological story differentiating them from certain other diseases. Still, these disease names were not senseless. In both cases, the epidemiological pictures made it plausible to assume that they were distinct disease entities. They had both appeared in salient, geographically and historically delimited epidemics. Therefore it is plausible to maintain that, when the etiological agents are not known, these disease names directly refer to hypothetical disease entities, which are unknown as to their intrinsic natures, but known only by their manifestations.

Malmgren thus concludes that when one knows that one deals with previously unknown diseases, one may name these diseases without being able to formulate definitions or even reliable clinical criteria for the single case. Finally, he extends his discussion to cover certain topics in psychiatric classification.

III. ON CAUSAL THINKING IN MEDICINE

The law of Causality, I believe, like much that passes among philosophers, is a relic of a bygone age, surviving like the monarchy, only because it is erroneously supposed to do no harm ([10], p. 180).

In spite of these celebrated lines by Bertrand Russell, causation remains one of the basic metaphysical problems in science. The causal mode of thinking continues, and the literature about causation grows in an exponential fashion. What then are the main issues on causation today, and how do these enter the Philosophy of Medicine? Let us mention four areas in the theory of causation which remain central, whatever science or whatever part of human affairs is concerned: (1) the characterization of the concept of 'cause' itself;

(2) causal analysis – how to distinguish causes from completely irrelevant factors; (3) the weighting between causes – the identification of the most important cause, and (4) the logic of causal explanation.

No contribution in this volume has (1) as its chief topic. Some contributors, however, inevitably touch upon problems of characterization. Henrik Wulff's essay even contains a substantial review of the late John Mackie's analysis of the concept of cause [7].

Two papers, Anders Ahlbom's completely and Anne Fagot's partially, focus on the second topic. They both discuss the problems of establishing and confirming causal hypotheses. Ahlbom treats hypotheses founded on data which consist of statistical correlations. Fagot deals with Bayesian probabilistic methods, in general.

The focal topic in this volume is the *third* one. Not less than six essays are, in their different ways, wholly or partly devoted to the problem of selection of causes and the weighting of causal factors.

The fourth topic, the nature of causal explanation, has attracted the attention of one contributor, Kazem Sadegh-zadeh.

IV. ON CAUSAL ANALYSIS

On what data are causal hypotheses founded? Given an effect E to be explained, and a background field consisting of events, A, B, C, how do we know that A is the cause and that B and C are irrelevant factors?

There are many ways to deal with such a question, and most of these have been thoroughly analysed in textbooks on scientific methodology. One of the *loci classici* in this literature is John Stuart Mill's *A System of Logic* [8], where Mill develops a theory of causal analysis. The methods developed by Mill all rely on the comparison between similar sequences of events where only one factor is modified at a time. By this process the irrelevant factors can be eliminated step by step. Consider the following simple instance of Mill's *method of difference*: A and B are followed by E. We remove B. As a result E disappears. Hence, B is the cause (or a non-redundant part of the cause) of E.

Mill's methods still remain useful in controllable situations. In complex human affairs, however, where controlled manipulation is difficult to achieve, or where manipulation does not give unequivocal results, other methods must be sought. Clinical medicine in general, and epidemiology in particular, nowadays rely heavily for their causal hypotheses on observed statistical correlations. The basic idea, then, is that, if a factor A is very often followed by a factor E (or if E is very often preceded by A), then it is quite likely that A is, in these cases, the cause of (or a part of the cause of) E.

Some theorists assume an even stronger link between statistical-probabilistic data and causality. For instance, in Patrick Suppes's work on causation [14] "*A* causes *E*" *means* that the probability of E given A is greater than the probability of E without A. (This is the basic idea; the complete theory contains many modifications.)

In his essay 'Criteria of Causal Association in Epidemiology', Ahlbom provides a survey of the problems facing the epidemiologist when he evaluates causal hypotheses by means of certain empirical data. One of the major problems in this kind of enterprise is to eliminate the possibility of so-called *confounding*. The relation between match-box carrying and lung cancer is a celebrated case of confounding. In spite of the high statistical correlation between these factors, they are not causally related.

Ahlbom discusses various criteria which have been proposed in order to tackle this and related problems. He pays attention to the criteria of biological plausibility, statistical strength, the dose-response relationship, and, in particular, quality aspects of the study design. Ahlbom's conclusion is that none of the mentioned criteria, nor any combination of them, is decisive. They can only have the status of plausible rules-of-thumb. The use of these criteria and the weighting between them must necessarily contain a great deal of subjective opinion.

In 'About Causation in Medicine', Anne Fagot makes similar observations. She discusses the limitations of a specific probabilistic method of causal analysis, the so-called Bayesian method. According to this method, we can make a rational decision between a number of causal hypotheses by the application of Bayes' theorem. The procedure, here illustrated with an example of etiological diagnosis, goes as follows: We look for the cause of a particular symptom in a patient. We have a limited number of diagnostic hypotheses. Possible causes are the diseases $D_1, D_2, \ldots, D_n$. Assume that we have some knowledge about these diseases, in particular their frequency and the likelihood of their giving rise to the symptom in question. These pieces of knowledge can be formulated in terms of the following *a priori* probabilities: $p(D_i)$, the probability of the disease, and $p(E/D_i)$, the probability of the symptom given the disease. By a repeated application of Bayes' theorem we can, then, for each disease calculate the *a posteriori* probability $p(D_i/E)$ – the probability of the disease, given the symptom. That disease for which the highest value is calculated in this procedure is considered to be the cause of the symptom. Hence we have the diagnosis.

This procedure for causal analysis with its obvious appeal to the mathe-matically-minded theorist has both theoretical and practical limitations. In illustrating some of the limitations, Fagot discusses a case where a physician tries to diagnose a patient with oscillating fever. A Bayesian calculation based on current medical knowledge and available background information about the patient suggests, as a resulting diagnosis, staphylococcal infection. This rational diagnosis, however, turns out to be false. The patient's wife eventually discovers the "real" cause. She notices that her husband's onsets of fever coincide with his taking antihypertensive pills. She makes him interrupt his drug intake and observes that this puts an end to the fever. Hence, by using Mill's method of difference, the wife provided the correct diagnosis.

One might have thought that this mistake was not due to any defect in the Bayesian procedure. The defect might lie in the background information. The Bayesian procedure, of course, presupposes adequate and exhaustive premises. Had anybody told the physician about the drug from the beginning, his diagnosis might have been different.

Fagot's critical point is that she doubts whether such additional informa-tion would have made the Bayesian diagnosis more accurate. The reason for this is that the drug in question does not normally give rise to high fever. The patient's reactions are quite specific. Therefore, some of the background data for the application of Bayes' theorem are such that the revised calculation would not speak in favour of the imbibed drug.

V. ON CAUSAL SELECTION

It is a commonplace in the current debate that every *effect* is preceded by a great number, perhaps an infinite number, of determining factors. We do not normally call all of these determining factors "causes". For instance, a common cold is determined by a multitude of events. There is not only the virus; we must also take into account such things as the climate, the constitution of the patient, and his general state of health before contact with the virus. We are, however, not prepared to call all these factors *the causes* of the infection. We normally select one or two of them as the causes. The rest would be collected under a heading such as "background conditions" or "the causal field".

A factor selected as *the* cause, then, is not a sufficient condition of its effect. It is only a part of a sufficient condition. This fact may be all right if we know what we do when we identify causes of phenomena and, in particular, if we use the same procedures in the selection. Many careful

studies indicate, however, that the criteria for *causal selection* are multifold
and that the use of these criteria may vary from situation to situation. We
sometimes select a condition because it is strictly necessary for its effect,
e.g., the tubercle bacillus for tuberculosis; we sometimes focus on an unusual
condition, e.g., hypothermia for a case of pneumonia; or we may sometimes
choose a particularly manipulable condition, e.g., smoking habits for a case
of bronchitis.

It is important to add that causal multiplicity is not one-dimensional.
There is not just one sufficient condition of a disease from which we select
one or two parts. There is a long, perhaps infinite, chain of sufficient conditions preceding the one which is most *proximate* to the effect in question.
Members of all of these are candidates for "causing" the same effect.

The variability of our causal thinking is not only of academic interest; it
has several practical consequences, not least in medicine. Some of these
consequences are explored in an illuminating way in this volume.

Fagot illustrates the importance of causal selection in the relation between
medicine and law. She discusses the case of a French woman who brought a
legal action against a doctor because he had failed in performing an abortion
operation on her. In court, the doctor was considered responsible for the
failure, since he had not taken all scientifically available steps to exclude
error. This decision of the court was highly criticized by some colleagues of
the doctor. They claimed that the "primary cause" of the young woman's
problems was not the negligence of the doctor but the fact that the woman
had had sexual intercourse!

Here *two* conditions are proffered as plausible circumstances necessary for
a particular effect. Which one is to be selected as *the* cause in order to assign
legal responsibility?

Staffan Norell discusses causal multiplicity from the point of view of
epidemiology. Unlike the legal case, however, epidemiology is not limited to
the monocausal idea. On the contrary, it is the duty of this discipline to
consider the whole field of determining factors. Norell illustrates this ambition
by describing some of the current causal models in epidemiology. In particular, the model called "the web of causation" attempts to give a fair
account of the causal complexity in both dimensions mentioned earlier.

On the other hand, the purpose of epidemiology is to provide knowledge
for the prevention of illness. Therefore, some kind of selection is eventually
called for. But the principle for this selection is evident: concentrate on such
determining factors the elemination of which will, to the greatest extent,
reduce the incidence rate of the disease in question. A model which might

help such a selection, and which is proposed by Norell, is called the "sufficient/component cause model". In identifying and studying various possible sufficient conditions for producing a particular disease, one may discover one element occurring in many or even all of these. The elimination of such an element will, of course, have a profound influence in the "battle" against the disease.

Ingemar Lindahl's 'On the Selection of Causes of Death' is related to epidemiological research. His topic is however quite specific. He analyzes the rules for causal selection formulated by the World Health Organization in the most recent revision of *The International Classification of Diseases, Injuries, and Causes of Death*.

Lindahl discerns and analyzes two kinds of conceptual problem in this context. First, there is a tension between the definition of the concept of '*the underlying cause of death*' (the term used for the primary cause to be selected in mortality statistics) and the specific criteria for selecting underlying causes. The definition restricts the application of the concept to one causal sequence and does not take into account the possibility of several *parallel* causal sequences. The rules for selecting the underlying cause, on the other hand, acknowledge the possibility of such parallel sequences. Secondly, the various criteria proposed are not always compatible with each other. For instance, the criterion of severity may clash with the criterion of prevention. The most serious factor (e.g., an advanced state of a lethal disease) is certainly not always the factor which is the most manipulable for the purpose of preventing death.

Causal analysis and the weighting of determining conditions play an important role in the identification and classification of disease. This is partly the topic of the essays 'Disease from a Historical and Social Point of View' by Øivind Larsen, and 'The Causal Basis of the Current Disease Classification' by Henrik Wulff. Both authors emphasize that the procedure by which we identify a disease entity is, in some important sense, arbitrary. Diseases, as we select them, are outcomes of long causal sequences of roughly the following kind: first, we have a set of external happenings occurring in a certain cultural and social setting; second, there is a set of bodily responses, sometimes initiating complex internal processes; thirdly, we have a set of consequences, some of an immediate, some of a remote kind, in terms of impairments and disabilities. The links that we recognize as diseases (and which are given labels in a disease taxonomy) can be found at all levels in this structure. But it has not always been like this. Wulff, who sketches the historical development, notes that before the 19th century, diseases were constituted by the last stages in these causal sequences, the actual suffering or disability of the patient. The first important reorientation came with the French school of pathology in the early 19th century. By then, diseases began to be identified as the *causes* of the patient's suffering. For instance, gastric ulcer was identified as the cause of many epigastric pains.

The search for the causes of diseases has run parallel to the nosological development. We discover (or better, *create*) new diseases by moving backwards in the causal chain as well as by turning to the microbiology of the human organism.

Larsen and Wulff would argue, however, that there is an additional important step to take in this development. The current classification is designed for those purposes which are the prevailing ones in Western medicine today, but which have been questioned more and more. Western medicine normally treats patients who are already afflicted with illness. Its purposes are therapeutic. Therefore, the entities to be classified must primarily be such links in the causal chain which are contained in the human body itself. If, on the other hand, medicine should have primarily preventive purposes, then medical attention and medical concept formation would have to be directed beyond the human body.

Larsen illustrates this point by describing a boy who regularly smoked 25 cigarettes a day. Epidemiological insights tell us that this boy runs a great risk of acquiring lung cancer in the course of the decades ahead. But although a causal chain involving disease is likely to have been initiated, there is no medical diagnosis applicable to this stage. The medical diagnosis is related to a late stage in the chain, viz, the eventual presence of a tumor in the boy's lungs. This is so in spite of the fact that prevention of the boy's suffering would have been much more effective and much less expensive had he been subject to some kind of medical intervention at an earlier time.

In looking for an efficient preventive illness program, Wulff discusses two kinds of causal analysis. According to the traditional method, one starts by observing an effect, a particular illness, and from there one reasons backwards in an attempt to identify causes among the various background conditions. In a medicine with illness-prevention inclination it might instead, says Wulff, be profitable to redirect the causal study. This means that one should start with a suspected "cause" and conduct prospective studies; study the various consequences of these "causes", given their particular environments. More concretely, this would mean the study of certain cohorts who are exposed to certain environmental conditions, be they of a physical, mental, or social kind.

An interesting and very general approach to the problem of causal selection in medicine is taken by Germund Hesslow in 'What is a Genetic Disease? On the Relative Importance of Causes'. The examples used in Hesslow's essay concern *causal selection* in the identification of disease. His analysis has, however, bearing on the weighting of causes in general. Hesslow's central thesis can be summarized in the following way: To say that X is a more important condition of Z than Y, is not to ascribe any intrinsic property to X; it involves, instead, a comparison between this case of causing Z and a reference-class of other cases of causing Z. Hesslow illustrates this point by considering the case of so-called genetic diseases. By a "genetic disease" is normally meant a disease whose most important determinant is a genetic factor. This is normally taken as some kind of absolute statement: there are genetic diseases and there are non-genetic diseases. Hesslow challenges this by showing that the disease, *lactose intolerance*, which is normally called genetic, appears as such only given a certain field of comparison. Given a Scandinavian population of reference, where milk consumption is standard and lactose deficiency is abnormal, it seems plausible to regard lactose intolerance as a *genetic* disease. But given an African population of reference, where milk consumption is rare and lactose deficiency is almost normal, the same condition would be looked upon as an *environmental* disease.

VI. ON THE LOGIC OF CAUSAL EXPLANATION

The philosophy of causal explanation has, during the last few decades, been completely dominated by a theory proposed in 1948 by Carl Hempel and Paul Oppenheim [6]. This theory has often been labelled the "deductive-nomological theory" of causal explanation. It has been given this title because of its two main features: (1) the idea that an explanation is a kind of deductive argument, and (2) the idea that an explanation depends upon the existence of universal laws.

According to the scheme proffered by Hempel and Oppenheim, the explanation of an event, for instance the incidence of a disease, amounts to giving a logical argument with statements describing one or more initial conditions and general laws as premises, from which the statement describing the event to be explained can be deduced. A simple illustration of such an explanatory scheme is the following:

Universal law – If sugar is put into water, then it dissolves.
Initial Condition – This lump of sugar is put into water.
Event to be explained – This lump of sugar dissolves.

This theory of explanation is very general. It is not limited to causal explanation. For an argument to be qualified as a causal explanation at least one further condition must be fulfilled: We must require that the universal law applied is a *law of succession*. This condition is fulfilled if and only if the antecedent of the law (the putting of sugar into water) is realized prior to the consequent's (the dissolution of sugar) being realized. The deductive-nomological scheme of explanation has become extremely influential in the philosophy of science. It has, however, also provoked powerful and sometimes devastating criticisms. The point of most of the criticisms is that Hempel-Oppenheim's criteria of explanation, in general, and of causal explanation in particular, are insufficient. There appear to be many arguments which fulfill the criteria of being deductive-nomological explanations, but which are nevertheless, from an intuitive point of view, inadequate as explanations. A common criticism is that the criteria do not prevent completely irrelevant laws and completely redundant factors from being used as explanans. A celebrated example is Salmon's (cited in Kazem Sadegh-zadeh's essay): "John Jones avoided becoming pregnant during the past year, for he has taken his wife's birth control pills regularly, and every man who regularly takes birth control pills avoids pregnancy" ([11], p. 205).

Sadegh-zadeh provides a lucid introduction to this subject in 'A Pragmatic Concept of Causal Explanation'. His main purpose, however, is to introduce a new scheme of causal explanation. This scheme still belongs to the tradition of Hempel and Oppenheim. Sadegh-zadeh retains the two fundamental cornerstones of the deductive-nomological explanation, the universal laws and the deducibility of the explanandum. However, he adds two important features to the basic scheme: the idea of *relevance* and the idea of *relativization* to language, time, and person.

For a universal law to be considered a causal law, Sadegh-zadeh requires that every element of the antecedent should be deterministically relevant for the consequent. This means that had any element of the antecedent been taken away, the "law" would no longer hold. (Salmon's law about men taking contraceptive pills would no longer count as a causal law.)

The feature of relativization is an original contribution by Sadegh-zadeh. He considers it unreasonable to believe that there are true causal explanations that every person should accept. A causal explanation is, therefore, only something which holds true relative to a particular person at a particular time: "... a set of statements which is a causal explanation for me may not be a causal explanation for you, and *vice versa*, if you and I subscribe to different systems of knowledge" ([11], p. 207).

VII. OTHER TOPICS ON CAUSALITY

The volume contains two further papers which relate problems of causality in medicine to other aspects of medical science. Erik Allander, in 'Holistic Medicine as a Method of Causal Explanation, Treatment, and Prevention in Clinical Work', analyzes and criticizes the role of so-called "holistic medicine" in framing and testing causal hypotheses in medicine. 'Holistic medicine' has, according to Allander, a very vague connotation. The term seems rather to refer to a movement of ideas which are only partially related to each other. Notions such as "wholeness", "cross-scientific research", "eco-philosophy" and "humanistic psychology" have a central place in this movement. Allander studied a number of works on medical issues which explicitly or implicitly relate themselves to the ideology of holistic medicine. In particular, he has studied some of their basic ideas concerning scientific research. Allander's conclusions are mainly negative: the advocates of holistic medicine want it to stay within the limits of conventional medicine without responding to acceptable scientific requirements concerning the framing and testing of scientific hypotheses.

In the final essay, 'Causes, Effects, and Side Effects: Choosing Between the Better and the Best', Stuart Spicker and H. Tristram Engelhardt, Jr. observe how much of the causal language in medicine is dependent on ethical and other evaluations. More specifically, the authors discuss how effects of medical intervention — be they effects of diagnosis, therapy, or care — are classified according to prevailing evaluative schemata. In particular, the distinction between *primary* effects and "*side* effects" of medical intervention is completely determined by the goal adopted for the intervention. The effect which is intended, i.e., which is identical with the goal, is the primary effect. The effects which are unintended are side effects. Variations in the purposes of a particular medical action will result in new discriminations between primary effects and side effects.

Causes and effects are then highlighted in terms of their usefulness in achieving [various] goals. Some effects are regarded as side effects, and some side effects as therapeutically unjustified side effects by appeals to these goals. This practice of medicine is certainly a rich interweaving of explanations and evaluations ([13], p. 231).

Linköping, Sweden,   LENNART NORDENFELT
May, 1983

BIBLIOGRAPHY

[1]   Aristotle: 1949, *Prior and Posterior Analytics*, ed. by Ross, W. D., Clarendon Press, Oxford.
[2]   Boorse, C.: 1977, 'Health as a Theoretical Concept', *Philosophy of Science* 44, 542–573.
[3]   Canguilhem, G.: 1978, *On the Normal and the Pathological*, D. Reidel Publishing Company, Dordrecht.
[4]   Engelhardt, H. T., Jr., and Erde, E. L.: 1980, 'Philosophy of Medicine', in Durbin, P. T. (ed.), *A Guide to the Culture of Science, Technology and Medicine,* The Free Press, New York, pp. 364–461.
[5]   Engelhardt, H. T., Jr.: 1984, 'Clinical Problems and the Concept of Disease', in this volume, pp. 27–41.
[6]   Hempel, C. G. and Oppenheim, P.: 1948, 'Studies in the Logic of Explanation', *Philosophy of Science* 15, 135–175.
[7]   Mackie, J.: 1973, *The Cement of the Universe: A Study of Causation*, Oxford University Press, Oxford.
[8]   Mill, J. S.: 1865, *A System of Logic: Ratiocinative and Inductive,* Longmans, Green, London.
[9]   Pörn, I.: 1984, 'An Equilibrium Model of Health', in this volume, pp. 3–9.
[10] Russell, B.: 1959, 'On the Notion of Cause', Chapter IX in *Mysticism and Logic and Other Essays*, 11th ed., George Allen and Unwin, London, pp. 180–208.
[11] Sadegh-zadeh, K.: 1984, 'A Pragmatic Concept of Causal Explanation', in this volume, pp. 201–209.
[12] Salmon, W. C.: 1970, 'Statistical Explanation', in Colodny, R. G. (ed.), *The Nature and Function of Scientific Theories*, University of Pittsburgh Press, Pittsburgh, pp. 173–231.
[13] Spicker, S. and Engelhardt, H. T., Jr.: 1984, 'Causes, Effects, and Side Effects: Choosing Between the Better and the Best', in this volume, pp. 225–233.
[14] Suppes, P.: 1970, *A Probabilistic Theory of Causality*, Acta Philosophica Fennica, Fasc. XXIV, Amsterdam.

SECTION I

ON THE CONCEPTS OF HEALTH AND DISEASE

INGMAR PÖRN

AN EQUILIBRIUM MODEL OF HEALTH *

I

In my contribution to this volume I try to tackle the characterization problem for health and some related issues. I am concerned to find a solution to this problem which is based on well-established distinctions and on models that have been shown to be viable. An important, relevant element of this sort is the distinction between systems of actuality and systems of ideality. In order to introduce this distinction let us follow Wertheimer ([8], pp. 88–92) and consider a law to the effect that anything which has a property P also has a property Q, and we assume next that a counter-instance to the law has been found, i.e., an object which has P but lacks Q. In this situation we can respond in one of two ways: either we blame the law and claim, with reference to the counter-instance, that the law is wrong or false, or else we blame the object concerned and claim, with reference to the law, that it is wrong or defective because it has P but lacks Q; either we criticize the law on the ground of the counter-instance, or else we criticize the counter-instance on the ground of the law. In the first case we work within a system of actuality, in the second case within a system of ideality. Examples abound. Theories in empirical science are systems of actuality, and so are ordinary maps and other pictures with real prototypes. Legal and moral systems, codes of etiquette, signalling systems, languages, building plans, volitions, desires, and interests are systems of ideality.

Judgements that a state of affairs, process, or change is normal or abnormal are relative to a system of ideality. In view of this I shall not use the distinction normal/abnormal in the context of systems of actuality — I speak instead of that which is average (in some sense) or deviates from the average.

An important class of systems of ideality is that of *standards of nature*. That is natural which exists in or by nature. When we speak of the natural color of one's hair or say that a person is curious by nature we use the word "natural" in that sense. A standard of nature is a system of ideality containing features or types (commonly termed ideal types) each one of which has instances in or by nature but not necessarily in such a way that any one object exemplifies or instantiates all of them. To see in the dark like a cat

3

L. *Nordenfelt and B.I.B. Lindahl (eds.), Health, Disease, and Causal Explanations in Medicine*, 3–9.
© 1984 *by D. Reidel Publishing Company.*

cannot be an ideal type in any standard of nature for human beings, but having an I.Q. of 140, two lungs, or an average pulse can be ideal types in such a standard.

Like other standards, standards of nature are standards of *evaluation*. Evaluation according to standards yields judgements to the effect that the thing or item evaluated is excellent, very good, fair, rather poor, very poor, and so on; and, as already indicated, evaluation according to standards yields judgements to the effect that the item evaluated is normal or abnormal.

Standards are relative to a point of view and instances of a given type (the reference type or class) may be evaluated from different points of view. Cars may be evaluated as means of family transport, as objects of social prestige, as sources of pollution, etc., and an individual car may at the same time be evaluated as an excellent object of social prestige, as a bad source of pollution, and as a poor means of family transport.[1]

II

Another concept central to this paper is that of goal-directed activity, a concept which is indispensable for the understanding of man as agent. In the study of goal-directed activity it has proved useful to employ the negative information-feedback control loop as the basic building block. An elementary loop of this kind comprises four units or stages. First, an actual inventory of goods of some kind, the volume or level of which is subject to change. Secondly, as information or inquiry system which registers the volume of goods in the inventory with varying degrees of accuracy. Thirdly, a decision process in which the actual inventory is compared to an ideal inventory, so called because it is a system of ideality that specifies what the actual inventory ought to be. A deviation by the actual inventory from the ideal inventory that is reflected in the information system is a discrepancy which must be removed or corrected, and in the decision process a decision is made as to how that is to be achieved from moment to moment. Fourthly, an implementation stage in which the task formed in the decision process is carried out, in part or in full, in an activity which affects the level of goods in the actual inventory and thereby influences the next decision – in so far as the change brought about in the activity is registered in the inquiry system.[2]

The ideal inventories that guide and direct an agent's activities belong to, so I shall say, his *profile of goals*. This profile also determines preferences among its constituent goals and thus it sets up dominance patterns which reflect the individuality of the agent concerned.

The ideal inventory of a loop is not generated within the same loop but imposed or supplied from without, but it may of course be generated within some other loop. The same is true of that collection of intrapersonal resources which I call the agent's *repertoire* and which is a feature of both the decision implementation stage and the inquiry system. The repertoire of an agent (at a given time) comprises his abilities (at that time), i.e., the things he can do in the sense of knowing how to do them or having the technique for doing them. The repertoire should be contrasted with the structure of external factors which we call the *opportunity* (for action). The repertoire and the opportunity are amalgamated in the agent's *powers to act* (at the time of the opportunity), i.e., the range of things which it is actually possible for him to do (at that time). This range determines the extent to which the agent's attempt to counteract the discrepancy between the actual and the ideal inventory succeeds or fails. And, likewise, it determines the speed, accuracy, and completeness of the inquiry system, considered as an activity which is designed to produce knowledge of the actual inventory.

The repertoire is relative to a set of psycho-physiological conditions. This is so because in a great many cases the abilities making up the repertoire are dependent on both mental and bodily conditions. A man without legs is unable to do a four-minute mile, but even if he retains the use of his legs he may never acquire the ability concerned because he lacks the required mental capacity. The ability to cross a street in moving traffic is obviously an amalgamation of bodily control and mental powers, and the same kind of amalgamation seems to be characteristic of great many abilities. By the *base* of a repertoire I shall understand the psycho-physiological conditions of its constituent abilities.

III

I am now in a position to formulate my dominant intuition about health: Health is the state of a person which obtains exactly when his repertoire is adequate relative to his profile of goals.

A person who is healthy in this sense carries with him the intrapersonal resources that are sufficient for what his goals require of *him*. This does not mean, however, that he will realize all of his goals, for his powers to act are determined not only by his repertoire but also by the external factors making up his opportunities for action — factors over which he does not always have control. Nor does it necessarily mean that, should the opportunities be forthcoming, he will realize his goals on his own, for some of the things he pursues may require of him, for example, the ability to cooperate with others or the ability to engage the support of others for his own ends. But it does mean that things other than his repertoire must be judged wrong or defective when his goals are not implemented.

Health, in the sense indicated, is genuinely relational in character. Neither the repertoire nor the goal profile is to be considered as fixed. The relation between them is one of mutual fitness: the repertoire fits the profile and the profile fits the repertoire. It is in this sense that a state of health may be described as an equilibrium. The point is perhaps best seen in the negative case, i.e., the case of illness, defined here as the contradictory opposite of health: the state of a person which obtains exactly when his repertoire is inadequate relative to his profile of goals. When a person is ill, in this sense, there are elements in his goal profile for the realization of which his repertoire is insufficient. The repertoire does not then fit the profile, but it may obviously be said, equivalently, that the profile does not fit, is not adjusted to, or is not aligned with the repertoire.[3]

IV

That health and illness are relations is a matter of some importance in, for example, the context of the distinction between illness and its causes, a distinction which, in my opinion, is crucial to the understanding of such categories as impairments, injuries, and diseases.

States, changes, and processes of an anatomical, physiological, or psychological kind may be evaluated according to different standards from different points of view. However, because of the fact that the repertoire has a psychophysiological base, it is certainly relevant to evaluate them from the point of view of health and illness — to judge them as factors which play a causal role in the maintenance of health and the production of illness. This point of view is implicit in the concepts of impairments, injuries, and diseases, or so it seems to me. I therefore think it is correct to characterize impairments, injuries, and diseases as, respectively, states, changes, and processes of an anatomical, physiological, or psychological kind which are evaluated as abnormal (poor, weak, etc.) because of their causal tendency to restrict repertoires and thereby compromise health.[4] It goes without saying that the standards involved are standards of nature; we are not prepared to term, say, some anatomical feature abnormal on the ground that it deprives the bearer of the ability to see in the dark like a cat or to fly like a bird.

However, it does not follow from these considerations that a person is ill if he sustains an injury, is affected with a disease, or is the bearer of an impairment. Owing to the relational character of illness and the nature of the relata, he is ill if and only if the injury, disease, or impairment makes *his* repertoire inadequate relative to *his* profile of goals.

Parallel reasoning may be advanced in the case of the goal profile. In this case, too, there are conditions which tend to cause a mismatch between repertoires and goal profile. Despair over oneself, which Kierkegaard calls a sickness of the soul, belongs to this category of conditions.[5]

The relational character of health and illness is of course also important in the context of health care and the treatment of illness. A person's ideality may demand things which can be delivered only by superhuman measures. It may be rare for a person to demand of his environment things that can be obtained only if he flies like a bird or swims like a fish, but less rare are cases of, say, aging persons who try to retain the powers of their youth in spite of the increasing weaknesses in their frames which make the attempts doomed to failure. In such a case the illness cannot be treated by measures designed to modify the repertoire; if health is to be maintained the profile of goals must be modified so as to make it fit the repertoire.

In the characterization of health and illness there is no reference to social or cultural factors. This is a consequence of the conception of health as an intrapersonal state of equilibrium. But social and cultural factors are relevant to matters of health and illness and frequently their relevance is of a causal nature. A state of illness may have a social origin. A person who is exposed to racial prejudice may be driven to resent the colour of his skin and to adopt goals which do not fit the repertoire. In cases of this sort the mismatch that constitutes an illness is caused by social factors, and any attempt to correct the mismatch that is based on respect for the individual must take this circumstance into account.

The characterization of health and illness is neutral as regards the epistemic access of the agent to his own health or illness, and, similarly, it leaves open the role of the agent's epistemic attitudes in processes affecting his health. I believe this to be correct, but I hasten to add that I also believe that these aspects are extremely relevant to the understanding of health in its full complexity. My reasons concern the stratification of the will. The will of a person may be stratified into systems of ideality of different orders. A person may want to have a will he does not have, or he may want to rid himself of a will he does have. He may also want to rid himself of the will to rid himself of a will he has, and so on.[6] Because of this kind of stratification the dynamic system of the information-feedback control loop may be given application at different levels of the ideality of a person. His profile of goals may thus be considered as the actual inventory of a second-order loop in which there is a decision process with an ideal goal profile, and an implementation stage and an inquiry system operating under the constraints of a repertoire in the form of abilities to modify the actual profile of goals and to acquire information about it. It is in the context of this repertoire and in the formation of the ideal goal profile that the agent's epistemic attitudes to his own health and illness play a major role. However, spelling out these matters in detail requires much more space than is available here. I therefore conclude by expressing the hope that I have said enough to indicate that the equilibrium model of health is pregnant with ramifications.

University of Helsinki,
Finland

NOTES

* I am indebted to Dr. Andrew Jones for helpful criticism of an earlier version of this paper.
1 For a thorough discussion of evaluation according to standards, see Taylor ([7], esp. Chs. 1 and 4).
2 For further details concerning this model of goal-directed activity, and for applications, see MacKay [3], Forrester [1], and Pörn ([5], Ch. 5).
3 In his paper [4], Dr. Lennart Nordenfelt mentions a solution to the characterization problem for health and illness which is similar, in some respects, to the one advocated here. Dr. Nordenfelt says of it that it avoids the conceptual circle he is concerned to present, but he finds it unacceptable for other reasons: "A final proposal which I would like to mention is the following: why not abstain from characterizing the vital purposes altogether and say that disability of unhealth is relative to the aims of the subject in question. . . . This idea is attractive since it avoids the circle altogether. It is, however, inadequate for other reasons. It seems unlikely that any aim, no matter how odd, that a person can have, should determine what his needs are. Some people have exaggerated goals; they aim for too much. We would never call those people medically disabled just for the reason that they fail in realizing such extreme goals" ([4], pp. 21–22). This is pertinent as a remark on medical disability or disease, but it is inconclusive as a reason against my own position since it requires the identification of illness with disease, an identification which is questionable and which, in fact, I go on to reject below.
   As regards the "circle of health" I wish to stress here that the solution I suggest imparts relevance, in the context of health and illness, to the entire goal profile and not merely to some select part of it (basic needs, vital purposes, etc.). As a result, the solution is not open to the charge of circularity of the kind Dr. Nordenfelt has in mind. In insisting on the relevance of the entire goal profile, I follow Whitbeck [10], a paper which I have found inspiring and admirable for its clarity. Her initial definition of health

comes close to my own characterization: "People generally recognize the value of having the psychophysiological capacity to act or respond appropriately in a wide variety of situations. By "apropriately" I mean in a way that is supportive of, or at least minimally destructive to, the agent's goals, projects, aspirations, and so forth. This good, I claim, is the good of health. The absence of any restrictions on a person's goals, projects, and aspirations in this definition is intentional" ([10], p. 611). Like Whitbeck, I believe it is important not to restrict the repertoire, and I think Dr. Nordenfelt's paper brings out the likely dangers connected with such restrictions.

4 Here, too, I have derived inspiration from Whitbeck, this time from her paper [9].

5 For a preliminary account of (elementary) despair, see my paper [6].

6 For elaboration and defense of the notion of the stratified will, see Frankfurt [2].

BIBLIOGRAPHY

[1]   Forrester, J. W.: 1961, *Industrial Dynamics*, The M.I.T. Press, Cambridge, Massachusetts.

[2]   Frankfurt, H. G.: 1971, 'Freedom of the Will and the Concept of a Person', *Journal of Philosophy* 68, 5–20.

[3]   MacKay, D. M.: 1956, 'Towards an Information-Flow Model of Human Behaviour', *British Journal of Psychology* 47, 30–43; reprinted in Buckley, W. (ed.): 1968, *Modern Systems Research for the Behavioral Scientist*, Aldine Publishing Company, Chicago.

[4]   Nordenfelt, L.: 1984, 'On the Circle of Health', in this volume, pp. 15–23.

[5]   Pörn, I.: 1977, *Action Theory and Social Science*, D. Reidel Publishing Company, Dordrecht, Holland.

[6]   Pörn, I.: 1981, 'On the Dialectic of the Soul: An Essay on Kierkegaard', in Pörn, I. (ed.): *Essays in Philosophical Analysis*, published as *Acta Philosophica Fennica* **32**.

[7]   Taylor, P. W.: 1961, *Normative Discourse*, Prentice-Hall, Inc., Englewood Cliffs, N. J.

[8]   Wertheimer, R.: 1972, *The Significance of Sense*, Cornell University Press, Ithaca and London.

[9]   Whitbeck, C.: 1978, 'Four Basic Concepts of Medical Science', *PSA* **1**, 210–222.

[10] Whitbeck, C.: 1981, 'A Theory of Health', in Caplan, A. L., Engelhardt, Jr., H. T., and McCartney, J. J. (eds.): *Concepts of Health and Disease*, Addison-Wesley Publishing Company, Reading, Massachusetts, pp. 611–626.

LENNART NORDENFELT

COMMENTS ON PÖRN'S
'AN EQUILIBRIUM MODEL OF HEALTH'

Professor Pörn's paper provides a very clear and elegant theory of health. It puts the notions of health, illness and disease into a well-developed theoretical framework, viz, systems theory, and it relates them in a clear fashion to other important anthropological notions, such as repertoire and profile of goal.

Pörn's contribution is in essential respects similar to another important theory of health recently proposed by Caroline Whitbeck ([3], [4]). In commenting on Pörn's paper I now wish to take the opportunity to compare Whitbeck's and Pörn's theories with respect to their basic notions. It could be enlightening to see to what extent Whitbeck's theory is vulnerable to the same objections as Pörn's.

Pörn and Whitbeck share the following general philosophy of health (with which I myself sympathize): Health is a molar concept pertaining to a person as an integrated whole. Instead of being a biological or biostatistical concept it should be viewed as an anthropological concept taking man as an agent in society.

More concretely, Pörn and Whitbeck have at least three important common platforms: (1) *Health is related to the subject's profile of goals.* Whitbeck expresses herself in the following way: "health is a person's psychophysiological capacity to act or respond appropriately (in a way that is supportive of the person's goals, projects, and aspirations) in a wide variety of situations" ([4], p. 620). (2) *Health and disease are not complementary concepts.* For Pörn the contradictory of health is illness and not disease. With Whitbeck there is no clear contradictory of health. Health is a dimension of which one can have much or little. A high degree of health is, as far as I can interpret Whitbeck, compatible with some degree of illness. (3) *Diseases, impairments, injuries etc. are psychophysiological processes, states, or changes which tend to compromise health.* For both Pörn and Whitbeck this statement is compatible with the fact that diseases etc. do not always restrict repertoires and thereby compromise health.

The most important differences between Pörn and Whitbeck, implicit in the previous statements, are the following ones: (1) With Pörn health is a *state of equilibrium* (the repertoire should be adequate to the profile of goals). Whitbeck has no such requirement. (2) As a consequence, health is

11

L. Nordenfelt and B.I.B. Lindahl (eds.), Health, Disease, and Causal Explanations in Medicine, 11–13.
© 1984 *by D. Reidel Publishing Company.*

an *absolute* term for Pörn. There can be no degrees of health. Only illness can be graded in terms of distance from the equilibrium. With Whitbeck we can have more or less of health.

My main objection to Pörn's theory is related to the equilibrium thesis. The crucial problems with this thesis arise from the variability of a person's profile of goals. If the goals are set too high, then no man, however strong and healthy, can reach them. And if they are set too low, then any man, however weak and sick, can attain them.

Consider first the possibility of a set of high goals. Pörn certainly realizes that there can be something wrong with the goal profile. He says, "A person's ideality may demand things which can be delivered only by superhuman measures" ([2], p. 7). In such a case, "if health is to be maintained the profile of goals must be modified so as to make it fit the repertoire" ([2], p. 7).

I agree that sometimes people may set goals for themselves which are so absurd and damaging to their lives that we wish to treat them, even medically, for having these goals. But this seems to be a rare case. What is not rare, however, is that people set more or less unrealizable goals. In these very common cases we have a mismatch between repertoire and goal. According to Pörn we then have illness.

I think it is obvious that this involves a considerable stretch of the common connotation of "illness". Should we accept the idea that all cases of *frustration* (when these have intrapersonal causes) are cases of illness?

Whitbeck's position is not so vulnerable in this respect, perhaps though, at the expense of clarity. She does not demand a complete adequacy between ability and goal. She requires only that the subject should be able to act in a supportive way with respect to his goals.

Whitbeck must, however, and does indeed, face a modified version of my objection. Sometimes, when the goals of a person are peculiar enough, his actions may not even be supportive of his goals. Will he then automatically be said to have a low degree of health? Whitbeck says, "It may be objected that some goals are themselves irrational . . . . To the extent that we believe that some goal is itself peculiar, however, or even irrational, we are inclined to say that it is something else that the person 'really wants' " ([4], p. 617).

Again I wish to say that not all unrealistic goals are irrational. We often miscalculate concerning our abilities without therefore being irrational or having irrational goals. More important, however, is Whitbeck's reliance upon the notion of a person's "real" goals. Her idea is that if we are inclined to apply the term "irrational" to a goal, then this goal can be seen to be intermediate to some further goal which is considered rational. The latter goal is the "real" goal ([4], p. 617).

It seems then that Whitbeck's main concern is with ultimate goals. The healthy man's actions should be supportive of his ultimate goals — via some adequate means. Without knowing more about the identification of such goals, however — for instance to what extent they are distinct from what I have labelled "basic needs" [1] — I shall not try to assess Whitbeck's standpoint in this respect.

So far I have queried the idea that all mismatches between repertoire and goal should be considered to be illness. I also wish to dispute that all equilibria should be considered healthy. Consider the cases where the goals set by an individual are extremely low; for instance, the mentally ill who happens to be content with doing nothing at all; or cases of long-term unconsciousness where there is no repertoire and no goals. These are indeed cases of equilibrium but hardly a healthy equilibrium.

As far as I can see Pörn has not explicitly anticipated this case. He has noticed that goals may be absurd and thereby issue in a mismatch. But he has not taken care of the case where goals, because of their absurdity, issue in an equilibrium. If I am right, then the fact of equilibrium cannot by itself be the touchstone of health.

Facing a similar objection — *mutatis mutandis* — Whitbeck's position is more promising. She can again resort to the real goals of the person. She can say that the real goal of the mentally ill is not to do nothing at all; and although the unconscious man cannot express any goals, he has some "real" goals. But again, Whitbeck has the obligation of explaining the notion of a "real" goal.

University of Linköping,
Sweden

BIBLIOGRAPHY

[1] Nordenfelt, L.: 1984, 'On the Circle of Health', in this volume, pp. 15–23.
[2] Pörn, I.: 1984, 'An Equilibrium Model of Health', in this volume, pp. 3–9.
[3] Whitbeck, C.: 1978, 'Four Basic Concepts of Medical Science', *PSA* 1, 210–222.
[4] Whitbeck, C.: 1981, 'A Theory of Health', in Caplan, A. L., Engelhardt, H. T., Jr., and McCartney J. J. (eds.): *Concepts of Health and Disease*, Addison-Wesley Publishing Company, Reading, Massachusetts, pp. 611–626.

LENNART NORDENFELT

ON THE CIRCLE OF HEALTH

The main issue in this paper is to reveal and discuss a conceptual circle that is embedded in some current hypotheses in the philosophy of health. I believe that this circle is worthwhile studying for at least two reasons. (1) It is not altogether trivial and obvious, since it involves more than one step of reasoning. (2) It is the result of quite plausible analyses of the concepts involved.

It should not perhaps come as a surprise that a basic and generic concept such as health invites a circular characterization. It may also in the end be true that the circle is unavoidable and that in a future theory some of the concepts involved have to be introduced jointly in the way Donald Davidson [3] envisages concerning the concept pair meaning-belief.

For the time being, however, I shall approach the circle from a traditional point of view: a conceptual circle is a problem and should be avoided in a philosophical reconstruction.

My procedure is roughly as follows: First, I simply present the circle and try to indicate why each step seems plausible. Second, I consider various strategies for dissolving the puzzle. And, finally, I elaborate somewhat on those alternatives which seem particularly promising. My results are not conclusive; I may, however, have indicated a fruitful direction of research.

HEALTH

The key concepts in the circle that I shall describe are 'health', 'ability', 'need' and 'health' again. I shall consider each concept in turn.

The definition of health is certainly highly controversial. A precondition for the circle to obtain is that we choose one of the alternative definitions. We should therefore ask:

What plausible characterizations of health are there? If we choose to reason on a sufficiently specific level we may certainly encounter an indefinite variety of alternatives. I cannot spell out here all such candidates. Instead, I shall propose the approach consisting in identifying some major types of health definition and indicate a few reasons for and against these.

In the current debate about health and disease there are, in fact, just two main dimensions which have attracted most of the interest. Let me

15

L. Nordenfelt and B.I.B. Lindahl (eds.), Health, Disease, and Causal Explanations in Medicine, 15–23.
© 1984 by D. Reidel Publishing Company.

label these the molecular/molar dimension and the non-evaluative/evaluative dimension.

The issue about the first dimension might be summarized in the following way: according to the molecularist, the criterion by which we judge whether a person is healthy or not is the normality of all (or almost all) molecular functions — be they physiological or biochemical. The state of the total person is in principle irrelevant. A person with a particular abnormality, say a cancer, is unhealthy however fit he feels and however much he can do.

For the molarist, on the other hand, the normality of the small functions are *per se* irrelevant. What really matters is whether the person as a whole is normal with respect to his vital functions or not. The molarist does not, of course, deny that there are molecular explanations of molar characteristics. What he does deny, though, is that there must be a determinate correlation between specific molecular functions and specific molar functions. He claims that certain abnormal functions on the molecular level are compatible with normal functions on the molar level, and vice versa. Accordingly, the sole study of the internal physiology of a person can never determine the issue of health.

So far we have left the concept of normality unanalysed. What kind of normality are we talking about here? There seem to be two main alternatives. Or perhaps again we should say that there are two main classes of alternatives. According to one alternative the matter of normality is determined ultimately by statistics; the normal is the most frequent, the average, or both. According to the other alternative the normal is that which is in accordance with some norm, in the sense that it is determined by a set of values; for instance, the individual's own values, the values of scientists, or the values of society in general. We thus see how our analysis of normality has brought us to the second of our two dimensions.

An application of the two dimensions gives us four basic views on the notion of health: (i) the molecular-statistical, (ii) the molecular-evaluative, (iii) the molar-statistical, and (iv) the molar-evaluative. Two of these positions stand out as the most favoured in practice. These are (i) and (iv). Good explanations of this fact can be given, but I shall not try to expound them here.

For my present purpose it suffices if I can make a case in general, for a molar conception of health. A defence of an evaluative and non-statistical molar conception will come later.

I think that the discussion of many writers (including Canguilhem [2], Goosens [5] and Engelhardt [4]) has shown that a molecular statistical conception of health is untenable. (This is not to deny that it can serve as a workable approximation in many cases.) The most salient objections show that many abnormal molecular functions are compatible with what we intuitively consider to be health. The subnormal pulse of an athlete is compatible with the highest degree of health. So is the transcendence of ordinary physiological functioning that an Indian Yogi can achieve.

Examples such as these show that we never, in our intuitive reasoning, consider molecular functions in isolation. (Our argument also holds, therefore, against a hypothetical molecular-evaluative point of view.) The low pulse of the athlete is well compensated by the strength of his coronary muscles. The end result, ultimately on the molar level, is therefore equivalent to (or even superior to) the end result of normal pulse in normal humans. The general reduction of many physiological functions that a Yogi can achieve is compensated by the extreme mental control that a trained Yogi maintains over his body. Certain ultimate functions, again on the molar level, are in his case not endangered.

Let this suffice for showing at least the plausibility of a molar conception of health. There are further considerations to be made in order to take care of certain objections. Particularly important is the objection that concerns early stages of such processes which we intuitively accept as diseases. Consider a person with a latent cancer, which has not yet resulted in molar consequences. Is a bearer of such a disease healthy or ill under the molar interpretation?

For answers to this and other objections I refer to the discussion in [9] and [10]. (See also the Introduction in this volume.) My central purpose here is not to work out a watertight conception of health.

I shall, therefore, accept from now on the molar platform and pursue the rest of my reasoning from there. Let us try to understand in more detail what is involved in the molar platform. What does it mean to say that a person as a whole is healthy, or to say that he is ill? There are two plausible general types of answer to this. One type of answer focuses on the feelings of the subject; the healthy man feels well; the sick man feels pain, fatigue or other feelings of discomfort. The other answer focuses on the abilities of the subject. The healthy man is able to perform a great many things; the sick man is incapacitated, handicapped, sometimes unable to do anything at all.

I think that both types of answer have a place in the characterization of health and disease. If one were to choose between the two approaches, however, I believe that the ability-type characterization would have a priority to the feeling-type characterization. Let me just mention two reasons here. (i) The ability-type characterization seems to single out all those cases that the

feeling-type singles out. The subject who is in pain, is tired or feels other dis-
comforts, is as a consequence of this normally also disabled. (ii) The feeling-
type characterization, however, does not single out all those cases that the
disability-type selects. The unconscious are totally disabled but do not have
any negative feelings. Certain mentally disturbed people are disabled but are
not consciously aware of any disturbance.

These considerations bring me to the conclusion that health should essentially be defined in terms of the subject's capabilities to perform certain
actions. Unhealth or disease should be defined in terms of disability or
handicap.

ABILITY

We have now reached the second step in our reasoning. We must ask: what
kind of ability and disability do we deal with here? What should a healthy
man be capable of doing? Or, conversely, what is a criterion of disability?
What is it that connects the blind man, the man with gangrene in his leg and
the psychotic man and justifies the common label "disabled"?

Here again there is the possibility of resorting to statistics. We can say that
the person who performs far below average in a number of activities is to be
called disabled. The blind, the immobile, and the psychotic clearly perform
less well than the average in many of the things they try to do.

This move won't do, however, for at least the following reason: Many
activities are irrelevant for a decision about health. The untrained clerk, who
is worse than most in his generation at playing football, is not automatically
disabled. The boy who cannot run 2000 meters without a break is not disabled just because most of his classmates can.

Instead, what seems to matter is rather whether or not the activity in
question is of importance, and in particular if it is of importance to the
subject himself. We ask whether it serves, or is needed for, some *vital purpose*
of his. Normally, football-playing and long-distance running are neither con-
tributory to nor necessary conditions for vital purposes. Non-ability to
perform such activities does not therefore normally count as a disability.

The assessment of a person's ability is, accordingly, related to some kind
of goal, which we have labelled a "vital purpose".

A person is disabled if he is unable to attain such goals, or to put the
matter more cautiously, if he is unable to fulfil certain necessary conditions
for achieving such goals. (We must keep in mind the possibility of certain
external preventative factors.)

Our task now is to determine the nature of vital purposes and their necessary conditions, i.e., the needs for these vital purposes. This brings us to a discussion of the concept of need.

NEEDS

The notion of need which is relevant for our purpose is a notion that has been given labels such as "basic need", "human need", or "vital need". (I have in mind such contributions to the theory of need as those of Braybrooke [1], Kaufman [6], Maslow [7] and McCloskey [8].) Let me here concentrate on the most well-known and comprehensive theory of needs, that of Maslow.

Maslow contends that needs have the nature of innate ("instinctoid") drives universal to all mankind. Every man — or (here comes the important qualification) every normal man — has a drive to satisfy his thirst or hunger, has a drive to achieve safety, to love and be loved, to acquire self-esteem, etc. All these drives indicate (or are, indeed, identical with) the basic needs.

In addition to this Maslow states: the fulfilment of these basic needs or drives is a precondition of health. "If a man has [actualized] basic needs in any active or chronic sense, he is simply an unhealthy man" ([7], p. 57). To paraphrase: if his basic needs remain unfulfilled for a long time, then the subject is or becomes ill.

A similar characterization is given by David Braybrooke: "deficiency in respect to [what he calls:] course-of-life-needs endangers the normal functioning of the subject of need, considered as a member of a natural species. In the case of men, such deficiencies might also be said to endanger health and sanity" ([1], pp. 90—91).

One could, of course, ask here: is this connection between the fulfilment of basic needs and health a connection *per definition*? Maslow does not clearly distinguish between conceptual and empirical characterizations. Can we not uniquely identify the basic needs without reference to health?

One possible procedure in the spirit of Maslow's is to try to identify those particular human drives that reflect the needs; to study people's behaviour and see what goals they as a matter of fact approach. Such an examination would, then, in principle, give us the list of basic needs without any appeal to health.

The trouble with this procedure, however, is that the basic drives are not the only drives that are or could be met with in human beings. In particular, certain sick people can have other drives, for instance, so-called neurotic drives. The behaviour that such people exhibit cannot help us in identifying basic needs or drives. In order to find the basic needs we must, therefore, first single out the class of healthy people. Hence, again we are confronted with the notion of health.

The circle that I announced at the beginning of this paper is thus closed. Health is defined as ability, ability means the power to fulfil basic needs, the fulfilment of basic needs is a necessary condition for health. In short: health is the ability to keep oneself healthy. (Or more cautiously: health is the ability to fulfil certain necessary conditions for health.)

I do not wish to draw exaggerated conclusions from this reasoning. All I have said so far is the following: If you make a series of hypotheses, which are at least initially plausible about the concepts of health, ability and need, you may end up with a definitional circle. I can now also add that some current influential definitions (which to my knowledge have not been considered together) point in the direction of such a circle. This, of course, does not imply that the circle is unavoidable.

There certainly are a number of strategies which present themselves. All of the notions involved, could be scrutinized again and assigned other characterizations. A statistical definition of health or a statistical definition of ability would save us from the circle. But is the price that we pay for this tolerable? Would we be correct in classifying all human statistical subnormalities of function as diseases or disabilities?

The most promising strategy is to scrutinize the notion of need. Is the ordinary notion of a basic need so tightly connected with health? There seems to be another workable proposal. The really basic needs are, of course, the ones whose fulfilment is necessary for mere *survival*. Survival is distinct from health and is also a biologically neutral concept.

This is, in fact, the approach chosen by the World Health Organization in its publication, *International Classification of Impairments, Disabilities, and Handicaps* (*ICIDH*). The disabled or handicapped man is, according to the *ICIDH*, one who cannot – in a satisfactory way – perform a *survival role* [11].

But, is this move sufficient for our purposes? Does not almost every list of basic human needs contain more than the conditions for mere survival?

There may be a more sophisticated interpretation of the strategy of survival, which would seem more plausible, and according to which survival is the ultimate purpose but according to which there are also restrictions laid upon the means for maintaining survival. Survival should be achieved *according to the expectations and regulations of society*. To be more concrete: what we, for instance, expect from an able and healthy man in a

Western industrialized society is that he can perform an ordinary professional role.

To fulfil a professional role is requisite in our society for satisfying our most basic physiological needs. Since professional life in Western society involves the performance of complicated actions, one is often obliged to enter into an involved series of actions in order merely to survive. (I am now talking about the standard requisite. Of course, most social security systems would prevent you from starving to death. But the point is that if society were to support you actively with extraordinary measures, then you would automatically be labelled a "disabled man").[1]

How should we evaluate this proposal? It is probably not entirely worthless as a piece of descriptive linguistics. It seems to mirror much current clinical thinking in the area. Ability to work is the most common criterion for physicians in determining whether a person is healthy or ill.

The proposed interpretation, however, seems only to save by accident most of the abilities which we intuitively grasp as criteria of health. Is there not a more plausible analytical characterization? I have no decisive answer here; let me just indicate two possible, although not completely adequate strategies.

I suggest first that we might link the concept of basic need to a more general concept than the concept of health itself. The general concept that I have in mind is the concept of well-being. Well-being certainly involves health but does not coincide with it. Other ingredients in well-being are ease, balance, happiness and a feeling of self-fulfilment.

A disease, according to this suggestion, negatively affects a person's ability to secure his well-being. For instance, if he doesn't fulfil his need for safety, friendship or love, then his well-being is endangered. But this need not entail that his health is threatened. He may become fearful or unhappy. These are not species of disease. Nor do they, as far as we know, necessarily lead to disease.

In making this move we have substituted well-being for health at the end of the conceptual string which initiated our discussion. This goes some way towards eliminating the circularity. However, as long as health is an ingredient in well-being, the conceptual circle is not completely removed.

A final proposal which I would like to mention is the following: why not abstain from characterizing the vital purposes altogether and say that disability of unhealth is relative to the aims of the subject in question. If a person wants to travel to New York, if he wants to repair his automobile, or if he wants to make friends with the president, and fails in performing these actions due to limitations which are internal to his body or mind, then he is disabled.

This idea is attractive since it avoids the circle altogether. It is, however, inadequate for other reasons. It seems unlikely that any aim, no matter how odd, that a person can have, should determine what his needs are. Some people have exaggerated goals; they aim for too much. We would never call those people medically disabled just for the reason that they fail in realizing such extreme goals.

A solution to our problem based on the aims of a particular subject must provide a way to eliminate extreme and peripheral goals, and be able to single out what we might intuitively call the real goals or interests of the person. The task of finding such a method cannot be pursued here.

CONCLUDING REMARKS

Let me sum up. My problem has been to find a way of avoiding a conceptual circle involving the notions of health, ability, and need. I have tried to show that this circle necessarily appears given certain plausible analyses of these concepts. In my attempt to break the circle I have considered a number of alternative strategies, all of them involving analyses of the concept of basic need. The link between basic needs and health seemed not to be self-evident. The first alternative related basic needs to mere survival, the second related needs to survival in Western society, the third related them to the generic state of well-being, and the fourth related them to the particular aims of a subject. I have discussed the merits and defects of these alternatives without, however, arriving at a final solution.

University of Linköping,
Sweden

NOTE

[1] Ordinary professional life entails, then, conceptually or causally, that many of those needs, which Maslow considers to be of a higher order, will also be fulfilled. In a curious way we may then have a reversal of the hierarchical order envisaged by Maslow. For most Western people a necessary condition for fulfilling one's physiological needs is the fulfilment of many of the higher order needs.

BIBLIOGRAPHY

[1] Braybrooke, D.: 1968, 'Let Needs Diminish That Preferences May Prosper', *American Philosophical Quarterly*, Monograph Series, Oxford, 86–107.

[2] Canguilhem, G.: 1978, *On the Normal and the Pathological*, D. Reidel Publishing Company, Dordrecht, Holland.

[3] Davidson, D.: 1974, 'Belief and the Basis of Meaning', *Synthese* 27, 309–323.

[4] Engelhardt, H. T., Jr.: 1981, 'Clinical Judgment', *Metamedicine* 2, 301–317.

[5] Goosens, W. K. I.: 1980, 'Values, Health, and Medicine', *Philosophy of Science* 47, 100–115.

[6] Kaufman, A. S.: 1971, 'Wants, Needs, and Liberalism', *Inquiry* 14, 191–212.

[7] Maslow, A. H.: 1970, *Motivation and Personality*, 2nd ed., Harper and Row, New York.

[8] McCloskey, H. J.: 1976, 'Human Needs, Rights and Political Values', *American Philosophical Quarterly* 13, 1–11.

[9] Pörn, I.: 1984, 'An Equilibrium Model of Health', in this volume, pp. 3–9.

[10] Whitbeck, C.: 1981, 'A Theory of Health', in Caplan, A. L., Engelhardt, H. T., Jr., and McCartney J. J., (eds.), *Concepts of Health and Disease*, Addison-Wesley Publishing Company, Reading, Massachusetts, pp. 611–626.

[11] WHO (World Health Organization): 1980, *International Classification of Impairments, Disabilities, and Handicaps (ICIDH)*, WHO, Geneva.

STUART F. SPICKER

COMMENTS ON NORDENFELT'S
'ON THE CIRCLE OF HEALTH'

The search for ideal definitions of concepts, the earlier master of which was, perhaps, Aristotle, has always appealed to philosophers whose quest for certainty was a guiding epistemological principle. In the last half-century, following the analysis of ordinary and "natural" language, a number of philosophers have concluded that there are in fact no neat or sharp boundaries that can, without exception, be drawn between concepts like 'disease', 'disability', 'handicap', 'deformity', 'normal' and 'healthy'.

Lennart Nordenfelt appears to pledge his allegiance to the Aristotelian tradition, and has set about to search for a singular definition of *health*, a difficult task at best and one which has led him to conclude (at least for now) that no final solution is quite forthcoming. It appears that in surveying some four notions of 'health', Nordenfelt has made a better case for a molar notion of health — one which strongly suggests that *the individual is capable of achieving a certain set of self-defining functions* in the straightforward sense of being able to do things called for in the everyday life-world, the world in which we all participate and in which we act.

Human capacity, very much socially defined, appears to set some boundary to any reasonable claim that a person is healthy. (But we should, of course, keep in mind the fact that any of us can at any time judge it good to be ill, since the notions of 'health' and 'illness' are clearly value-laden notions.) So whether or not the practice of medicine — I mean clinical medicine — enables patients to recover, it is surely possible to locate a variety or *family* of notions of health running, so to speak, below the everyday surface and usage, and Nordenfelt has cited a few of these. But I cannot help noticing here the obvious absence of the patient's view of the meaning of 'ability', 'capability', 'functioning', and 'needs'. Such *subjective* assessments appear crucial to any coherent view of health; nevertheless Nordenfelt is no doubt correct that the person's *feelings* are less relevant than his or her *capacities*. The difficultly with any attempt to reduce all notions of health to a singular, essential and all-encompassing (or ideal) definition is that such a definition as the *molecular* one seems to have a compelling force and use *only* when we have some idea of the specific purposes, aims and intentions of (say) the pathologist. Can we, after all, really say that a malignant tumor growing in a person who is

25

L. Nordenfelt and B.I.B. Lindahl (eds.), Health, Disease, and Causal Explanations in Medicine, 25–26.
© 1984 *by D. Reidel Publishing Company.*

unaware of it (indeed, who feels quite well) is not a contributor to ill-health and disease? (Alvan Feinstein's important remark, in *Clinical Judgment*, on "lanthanic disease"[1] surely has an important place in our notion of disease.) If we try to make light of the carcinoma, we simply choose to ignore one model of the concept of illness: in all likelihood it will become critical at a later time, for example, just before surgery and/or chemotherapy.

I therefore ask Nordenfelt two questions: First, can he not accept the notion of a "family" of concepts of health, where each one is acceptable to the degree to which it serves a particular set of purposes and uses? Being a practical and applied science, medicine does not seem to lend itself to singular notions or definitions of 'health' and 'disease', in spite of our deepest hope that with Aristotelian verve we can locate such an all-encompassing definition of these notions.

Secondly, and finally, I ask: Is the connection of concepts − health → ability → need → health − the formation of a vicious circle, as Nordenfelt has presented it? Is this "movement" of concepts incoherent? Are there tacitly hidden in this "circle" *two* competing and conflicting concepts of health? Again, is the circle really coherent as described; is there really a circle at all? A careful analysis of this set of concepts is still in order, for Nordenfelt's own prejudices do manifest themselves: He clearly holds that health is *properly defined* in a molar way; on this point I am in disagreement with him. For 'health' is a family of concepts and therefore it is far better to ask after the purposes of the pathologist, the internist, the patient, the employer of the patient, the nurse, and the philosopher . . . .

University of Connecticut School of Medicine,
Farmington, Connecticut, U.S.A.

NOTE

[1] Feinstein, A.: 1967, *Clinical Judgment*, Robert E. Krieger Publishing, Co., Huntington, N.Y., p. 145.

H. TRISTRAM ENGELHARDT, JR.

CLINICAL PROBLEMS AND THE CONCEPT OF DISEASE

Language itself can misguide, and concepts by their very character can mis-direct us. Disease concepts have suffered from the assumption that they name things in the world in a value-free fashion. They are instead goal-directed notions. The history of medicine shows some of the roots of this confusion in the process of transforming clinical problems into disease entities. Even ancient medicine offers such puzzles concerning the nature of concepts of health and disease. Indeed, there is some evidence that the physicians of Cos and Cnidos were in disagreement concerning what should count as a single disease entity. The physicians at Cnidos, it would appear, held that even minor differences in symptoms justified the creation of a new classificatory unit ([25], p. xiii). These disputes set the stage for a long history of conten-tion with regard to whether diseases have a reality in and of themselves, or whether they are the creations of physicians. Views of the first sort have supported the notion that diseases are in some sense beings or entities, *entia morborum*. Individuals supporting such understandings of disease have been termed ontologists.[1] One finds, for example, François Broussais (1772–1838) criticizing rival accounts of disease as ontologies ([7], Vol. 2, p. 646). The non-ontological views have, for their part, been variously termed functional or physiological accounts of disease [43]. These have often taken a frankly nominalist approach to the status of disease entities ([13], [14]) by recog-nizing only individuals as real and holding sicknesses or diseases to be con-structs properly fashioned with reference to their instrumental value in making reliable predictions, identifying regularities in nature, and directing successful therapeutic interventions.[2] This contrast between ontological and physiological accounts has been explored in the 20th century by Knud Faber [19] and Richard Koch [29].

As accounts of the history of concepts of disease indicate, there is a complex interplay of understandings of the nature of disease ([10], [11], [15], [32]). In particular, there is not one ontological understanding of disease. Disease entities have in part been understood in etiological terms, as causally efficacious entities. Here one might think of accounts of disease ranging from those of Paracelsus to those of the bacteriologists of the 19th century [33]. They have also been construed in patho-anatomical or patho-

27

L. Nordenfelt and B.I.B. Lindahl (eds.), Health, Disease, and Causal Explanations in Medicine, 27–41.
© 1984 *by D. Reidel Publishing Company.*

physiological terms. Here in fact, Broussais, who criticized the clinical ontologists, would appear to be an ontologist despite himself. It is simply that
for him disease entities are patho-anatomical entities ([8], p. ix). Indeed,
among those who were at least in some parts of their careers ontologists in
this sense, one would need to list Rudolf Virchow (1831–1902) [46], who
went from being anti-ontological ([45], p. 188) to having a pro-ontological
viewpoint ([46], pp. 190–195). Finally, there is a rich history of clinical
nosologists who held that constellations of signs and symptoms could be
identified in natural and enduring patterns. Among these would be Thomas
Sydenham (1624–1689) [41], François Boissier de Sauvages (1707–1767)
[39], Carl von Linnaeus (1707–1778) [30], and William Cullen (1710–
1790) [14]. In short, ontological understandings of concepts of disease were
fashioned to support diverse understandings of medicine [42].

Though one might construe the debates concerning the nosologies of the
18th century as arguments regarding the extent to which classifications of
diseases are natural or artificial, one might also understand them as disputes
regarding the most useful ways in which to approach medical reality, where
"usefulness" is assessed not only in terms of the capacity to further develop
explanations of disease, but in terms of useful treatments for patients as well.
It is thus tempting to consider the shift from the clinical classifications of
Sauvages and Cullen to the patho-anatomical approaches of Xavier Bichat
(1771–1802) [2], Broussais, and Virchow, not so much as an element in
the ontological quest for the true reality of disease, but as an advance in the
search for models and modes of examining medical phenomena that could
provide more encompassing explanations of disease phenomena, and in the
end provide for more useful forms of treatment.

This historical record can support both accounts. Physicians of the 19th
century undoubtedly had quasi-metaphysical goals in mind. However, in that
medicine is always tied intimately to therapeutic goals, a revisionary interpretation suggesting an instrumentalist understanding of medical generalization
would appear to be in accord with the way in which medicine is usually
practiced. One can at least find hints of this understanding in some of the
nosologies of the 18th century. One might think here of Sauvages, who in
the last edition of his *Nosologia methodica* devoted over 80 of the 1500 pages
to etiological and anatomical classifications of disease. He offered alternative
ways of construing diseases. Further, though Sauvages appears to have held
that species of disease are discovered, he recognized that the genera and their
arrangements were often imposed by the nosographer ([39], Vol. I, Section
100, p. 28). Also, Sauvages, in not dismissing concerns with etiological or

anatomical accounts, underscored their usefulness, though he appears to have held that such concerns would not disclose the true essence of diseases. In fact, his argument that one ought to employ clinical classifications appears in part to turn on considerations of which approaches to medical reality will be most trustworthy clinically, and therefore most useful, not which are most likely to disclose medical reality *sub specie aeternitatis* ([40], Vol. I, Section 73, p. 21).

In addition to the debates concerning the ontological status of concepts of disease, an interest has developed in determining the extent to which concepts of disease are value-dependent, and the extent to which such values may be dependent or independent of the particular cultures in which they have been elaborated ([2], [3], [4], [5], [6], [16], [18], [26], [27]). Individuals such as Christopher Boorse have argued that a concept of disease can be developed in a value-free fashion by reference to ". . . a type of internal state which is either an impairment of normal functional ability, i.e., a reduction of one or more functional abilities below typical efficiency, or a limitation on functional ability caused by environmental agents" ([6], p. 567). In contrast, Boorse holds that illnesses reflect societal values: "Being ill involves having a disease serious enough to be somewhat incapacitating, which thereby supports normative judgments about treatment and responsibility" ([6], pp. 551–552). Others have joined Boorse in defending value-free or at least culture-independent concepts of disease ([26]; [47], pp. 53–54). The difficulty lies in identifying a truly value-neutral standard. As Boorse himself acknowledges, his approach to identifying concepts of disease in terms of departures from species-typical levels of species-typical functions makes it difficult for him to denominate universal "diseases" as diseases. Thus, widespread phenomena such as arteriosclerosis, and genetically predetermined decrements of physiological function connected with aging, cannot easily count as diseases in Boorse's terms. Further, as William Goosens has shown, Boorse's argument becomes peculiarly dependent upon the past deliverances of evolution. Current species-typical levels of species-typical functions reflect adaptations to environments in which humans no longer, for the most part, live [22].

The difficulty with Boorse's position is that it will either describe a state of affairs of interest to taxonomists classifying the peculiarities of particular species, and not be equivalent to the concepts of disease employed by physicians, or it will indeed require an appeal to particular values. Thus, in defining normal function as "a part or process within members of [a] reference class [which part or process] is a statistically typical contribution by it to their individual survival and reproduction" ([6], p. 555), Boorse ties his concepts of disease and health (or at least non-positive concept of health) to a reference point which would not ordinarily be central to clinical medical considerations. Clinical medicine is not focused on states of affairs as diseases because they do not maximize reproductive capacities to a species-typical level. If that were the case, one could face the possibility of characterizations that would appear to be counter-intuitive. For example, if having an I.Q. greater than two standard deviations from the norm leads to less reproductive success than that of individuals within two standard deviations of the norm, it would count as a disease state. The difficulties with attempts such as Boorse's become clearer if one attends to the ambiguities in concepts of normality. As Marjorie Grene has indicated, there are at least three major notions of normality employed in the language of bio-medicine. One sense is synonymous with health and is value-laden. The second is synonymous with being statistically frequent. The third is of special interest to taxonomists and means something like "characteristic for members of the species" [23].

In fact, if one examines the problems of defining health within a post-Darwinian understanding of biology, one sees that it will not be possible in fact to have one sense of health or of disease. Instead, one will find various families of concepts. If one attempts to understand health as a successful adaptation to a particular environment, one will need to specify not only the environment, but the goals for success. That is, once one does not presume that humans were created as adapted to some putative ideal environment (e.g., the Garden of Eden), one will need to specify the environment in question. A dimension of such an environment will be the culture in which the humans in question live. Further, it is not clear what the goals of adaptation ought to be. They will quite clearly differ if one is interested in maximizing individual reproductive success, inclusive reproductive fitness, or individual fulfillment. A mere study of how individuals have responded to past selective pressures will not disclose to a disinterested observer what rational individuals should seek as the goals of their adaptation. At best, one will be able to indicate ranges of goods in which individuals tend to have interest. And insofar as one is concerned with developing concepts of disease or health within the domain of medicine, one will be interested in determining which goals medicine can aid in achieving. Thus, disease concepts will involve reference to problems in achieving expected abilities to function,[3] freedom from pain, and expected human grace and form, insofar as these problems are seen to be physiologically or psychologically based and beyond the direct and immediate will of the individual whose problem it is [17]. Since certain physiological and anatomical states are likely to impede the achievement of important human goals in any environment in which individuals are likely to live, they will constitute cross-culturally recognizable diseases [3]. However, states of affairs such as color-blindness and menopause will in some circumstances count as diseases or defects, and in others they will not. The judgment in these matters will be properly dependent upon the environment, and upon what individuals in particular cultures will accept as proper expectations ([28], [49]). For instance, color-blind individuals can identify camouflage with greater success than color-visual individuals ([35], pp. 131–146). Here, one might imagine an environment where the capacity to spot camouflage either maximizes one's own chances to reproduce, or maximizes inclusive fitness.

In short, what medicine addresses as diseases is a cluster of physiologically or psychologically based problems with function, freedom from pain, and bodily form. They are identified as problems because they blunt the realization of particular human goals. They are problems for medicine not because they are species-atypical levels of species-typical functions. Problems become medical problems because they bother individuals, or bother others about individuals, and can be ameliorated or accounted for by medicine. In short, drawing upon Marjorie Grene's suggestion that interests in what is normal in the sense of species-typical are more pertinent to taxonomists, one must suspect that those who have undertaken endeavors such as Christopher Boorse's have in fact, if they have succeeded, reconstructed a concept of disease which is not directly relevant to clinical medicine. Since medicine can address species-typical decrements of species-typical functions as diseases, as for example those associated with aging, it follows that Boorse's characterization of diseases is not that of a necessary condition for a state of affairs counting as a disease for clinical medicine. Boorse himself has in fact admitted this much [6]. Nor need species-atypical levels of species-typical functions that would lead individuals to have a decrement of reproductive capacity below species-typical abilities count as diseases, as may be the case for very high I.Q. levels in particular environments. Boorse has not characterized a sufficient condition for a state of affairs being held to be a disease for clinical medicine.

Further, one should note that evolution selects for inclusive fitness, not simply the reproductive fitness of each individual [48]. Therefore, evolution will produce kinds of individuals who do not themselves reproduce, but who in the past maximized the reproductive advantage of the group of which they were a part. One might think here of worker bees. Thus, contrary to Boorse, homosexuals may not be diseased [3], given an evolutionary perspective, if the trait developed to maximize fitness, as has been argued by some [44]. In short, Boorse may not have even succeeded in providing an adequate reconstruction of a concept of "disease" as it might be employed by an evolutionary biologist.

It is therefore useful to reassess the pursuit of a reconstruction of the concept of disease as it is employed in clinical medicine. One ought to recognize that natural languages offer a number of allied terms such as 'disease', 'defect', 'deformity', 'disfigurement', or 'malady', with poorly defined meanings and poorly defined boundaries. They are of interest to medicine insofar as they constitute actual or potential clinical problems. Here a review of the nosologies of the 18th century, in particular those of Sauvages [40], can again be useful. One finds Sauvages listing an inventory of clinical problems. Medicine has traditionally started from clusters of symptoms and other findings which one might stipulatively term 'illnesses' (including defects, deformities, etc.), and then moved to provide, as far as possible, disease explanations. The search for species-typical deviations of physiological or anatomical functions or structure often serves as a good heuristic device, though as indicated above, such will not be a sufficient guide. Such conditions must also violate certain general expectations regarding desired function, freedom from pain, and human form. Further, since medicine is a social enterprise, generalizations concerning illnesses, and explanatory accounts of them, are harnessed to particular therapy roles into which patients are cast. Thus, there are descriptive, explanatory, and performative elements of medical disease language [18].

Evaluations play roles in different ways in all of these elements. Clusters of findings exist as explananda for medicine because they cause states of affairs held in some sense to be improper [16]. Patho-physiology is distinguished from physiology as being the physiology of some variety of suffering. Further, since clinical medicine must organize data so that physicians can employ them, explanations are forwarded in ways that often reflect views regarding what the most prudent or cost-effective understandings of such data are. Thus, tuberculosis is likely to be characterized as an infectious disease by internists, even though individuals working in social medicine may see it as a disease tied to poverty, or those working in genetics as a disease that may have important hereditary components [15]. Finally, the decision as to when a particular patient ought to be placed within a particular therapy role should reflect costs of over- and under-treatment, both for that individual and for society. This is to suggest that most disease language expresses social strategies for coming to terms with physiologically, anatomically, or psychologically based problems with function, pain, or human form through the application of the various powers of medicine. The development of classifications of disease, though they are in part environment- and culture-dependent, and express particular goals within particular cultures, are not arbitrary within any particular societal or cultural context. Still, there will not be the possibility to elaborate either univocal or value-neutral, culture-free concepts of disease in that one will not be able to identify *the* paradigmatic environment or an absolute set of goals for human function and human form. Further, individuals will typically be able to exempt themselves from certain elements of therapy roles, especially when the disease involved is not contagious. One might think here of the recent example of the taxon 'ego-dystonic homosexuality' created by the American Psychiatric Association. Such a characterization on the one hand allows homosexuals seeking treatment to be placed in a sick role, often permitting them to have their care reimbursed, while on the other hand recognizing that when individuals do not derive difficulties from their sexual orientation, it ceases to be sensible to term it a "disease" ([1], p. 281). Clinical medicine comes to address a variety of problems that are to be appreciated fully as clinical problems only within the context of particular environments and particular goals, both individual and societal.[4]

In short, there may be a major error in seeking *the* nature of disease as it is presupposed by clinical medicine. Clinical medicine, it would appear, does not presuppose *a* single notion of disease. Clinical medicine is focused on resolving clinical problems: problems with pains, expected human form and grace, and expected abilities to function, insofar as those problems are seen as having a physiological or psychological basis, and as being beyond the direct and immediate volition of the individuals held to "have" the problems [17]. Thus, medicine addresses with equal propriety as clinical problems, viral hepatitis, schizophrenia, the pains of teething and of childbirth, unwanted fecundity and undesired sterility, as well as difficulties such as appendicitis. Moreover, medicine has traditionally addressed this wide range of problems.[5] One might, therefore, wish to revise slightly the excellent argument of Henrik Wulff that, "A disease entity may, therefore, be regarded as the vehicle of clinical knowledge and experience . . ." ([50], p. 68). One can usefully substitute "clinical problem" for "disease entity". In this fashion one can escape the metaphysical quest of individuals such as Clouser, Culver, and Gert, to discover the nature of disease entities or maladies.[6] One can instead focus on examining the role of various characterizations of clinical problems in clinical medicine which, as Wulff has argued, employ disease concepts in terms of their usefulness, not so much in terms of their truth *sub specie aeternitatis* ([50], p. 69). Classifications of clinical problems provide "mental resting places for prognostic considerations and therapeutic decisions . . ." ([50], p. 80). The interests of clinicians are pragmatic and instrumental.

Further, since the goals addressed by clinical medicine are diverse, it would likely not be useful to seek one sense of clinical problem or of health [9]. The focus of clinical medicine is diverse because the well-being of humans is diverse. *A* concept of health or of disease is unlikely to be of use in clinical decision strategies which require the weighing of data according to various rules for maximizing particular types of well-being and minimizing particular sorts of morbidity, financial costs, and mortality risks. When one examines, for example, the excellent reconstruction of the process of clinical judgment by Kazem Sadegh-zadeh, one finds little reference to diseases, but rather to rules under which data can become reliable warrants for therapeutic interventions which achieve the interests of physicians and patients (presumably to preserve, or as far as possible to restore, the ability of individuals to function, to be free of pain and to have desirable human form).[7] Clinical medicine is interested in reliable warrants for useful medical interventions.

My point has been here to suggest that for an analysis of clinical medicine, as opposed to the basic medical sciences, it is best to substitute concepts of clinical problems for traditional concepts of disease, and to see the characterization of clinical problems as a step in clinical decision-making. That is, characterizations of clinical problems are ways in which one transforms ill-structured problems into problems with a structure useful with regard to offering predictions and other therapeutic interventions. Even predictions and prognoses are offered in clinical medicine as elements in the decision processes of patients and physicians with regard to exposures to particular risks. They are made against special concerns with falsely over- and under-predicting particular outcomes. Here one should note that the meaning of scientific terms is best found in the processes in which they are employed. As a consequence, one will need to distinguish between concepts of disease as they exist in unapplied "medical" sciences versus applied medical sciences. Concepts of disease as they might be employed by evolutionary biologists to indicate the reduction of the inclusive fitness of an individual are thus to be contrasted with concepts of disease as they are employed by physicians, as warrants for particular kinds of treatments or prognoses being given to patients. In the first case one finds "diseases" as elements of a process of explanation. In the second case one finds "diseases" as elements of a process of intervention. Clinical problems are problems for explanation within clinical medicine insofar as better explanatory models are useful in the achievement of well-being or avoidance of impairments.

Clinical medicine is, in short, an applied science, an applied systematic body of knowledge. It is not a search for explanations and predictions for their own sake, or on behalf of a better understanding of reality. Rather it is a search for explanations and predictions in the service of non-epistemic goals such as the achievement of well-being or the avoidance of impairments. Since one is interested in applying knowledge to non-epistemic goals, the success of such applications is judged primarily by non-epistemic standards.

When one recognizes clinical problems as points in a decision-making process focused on offering useful therapeutic interventions (including preventive medical interventions) and prognoses, it not only becomes clearer why clinical problems rather than classical notions of diseases or disease entities should be more the focus of philosophical inquiry concerning clinical medicine, but also how they are bound to notions of causation in clinical medicine. Ideally, clinical problems should be characterized in ways to suggest appropriate therapeutic interventions. Again, since clinical medicine is an applied science, there will not be an interest in causality as it occurs in non-applied sciences. Rather, there will be an interest in theories of causation, somewhat analogous to the approach of the law. One will wish to isolate from a complex of causally relevant factors those that ought to be held therapeutically accountable. One will wish to identify not all causal factors, but those that are important in the prevention, treatment, or cure of clinical problems, and in making useful predictions (prognoses) for physicians and patients. As I suggest in my commentary on Wulff's paper in this volume, one might consider Hart and Honoré's account of how the law determines which causally efficacious actions and omissions ought in fact to be held to be the *causes* of an outcome, and which ought to be simply regarded as *background conditions*. One might think of their example of the gardener who is held responsible for flowers dying in a garden he ought to have tended, though the lack of rain and the failure of passers-by to water the garden are, apart from the interests of the law's concern with responsibility and accountability, equally causally related to the outcome ([24], p. 24). The language of diseases, or better still, clinical problems, is thus placed within a pragmatic account of causation in which those factors which are usefully held to be accountable in therapeutic practices are held to be *the* causes. There is no harm in referring to some causes of clinical problems as diseases as long as one does not forget the pragmatic context of causal language in medicine.

When clinical problems as seen from the point of view of patients are related to patho-physiological, patho-anatomical, or patho-psychological accounts, there is already a selection of what to accent as causal factors, in terms of the useful therapy roles available for individuals with such problems ([18], [50]). Thus, phenylketonuria is likely to be seen as a genetic clinical problem or a metabolic clinical problem, depending on whether one is providing genetic counseling to prospective parents who carry the recessive genes for the trait, or whether one is explaining to parents the reasons for providing a phenylalanine-free diet to their newborn child. Clinical categories are embedded in various clinical problem contexts, which alternately cast different causal factors as background conditions or as causes. If one were seeking to understand diseases as entities existing apart from problems of clinical decision-making, this point would not be as obvious. As one comes to see the characterization of clinical problems as properly tied to the goals of therapy, and of medical intervention generally, one should see as well why "disease entities" assume a pragmatic character. They are properly fashioned as elements of strategies to be employed in solving problems in clinical decision-making. Clinical medicine is not developed in order to catalogue diseases *sub specie aeternitatis*, but in order for physicians to be able to make more cost-effective decisions with respect to considerations of morbidity, financial issues, and mortality risks, so as to achieve various goals of physiologically and psychologically based well-being. Thus, clinical categories, which are characterized in terms of various warrants or indications for making a diagnosis, are at once tied to the likely possibilities of useful treatments and the severity of the conditions suspected. The employment of a particular taxon in a classification of clinical problems presupposes, as a result, a prudential judgment with respect to the consequences of being right or wrong in the circumstances [18]. Such appraisals of consequences in turn are made in terms of appraisals of causal factors, where the accent is laid upon the factors that can be manipulated, and therefore held to be accountable.

In summary, recent reappraisals of the nature of disease and of the character of medical decision theory offer a new perspective upon the traditional disputes regarding the standing of disease entities ([18], [21], [37], [50]). First and foremost, they suggest that the character of traditional discussions is misleading: it supports a reification of disease entities. Instead, what is required is an analysis of the role of classifications of clinical problems in medical decision theory which appreciates clinical problems as elements in strategies for structuring ill-structured problems with respect to the goal of therapeutic interventions. Talking of "clinical problems" rather than "diseases" can then function as a linguistic reformulation aimed at freeing analyses from old metaphysical associations. It would be too facile to say that this warrants a triumph of the anti-ontologists in clinical problem solving. Rather, important implicit elements in clinical decision-making have been made explicit and the role of particular goals and values in medical decision-making can now be forthrightly addressed. Moreover, these concerns are most congenially lodged within pragmatic accounts of causation where particular causal factors are isolated as *the* causes because of the usefulness of that isolation in therapeutic practices. It is to the merit of much of the current work regarding clinical judgment and medical decision theory that traditional debates concerning the nature of clinical problems are being recast in new and most likely more fruitful terms.

Baylor College of Medicine,
Houston, Texas, U.S.A.

NOTES

[1] The term 'ontological' here should be regarded as one peculiar to the history and historiography of medicine, with only distant consanguinities with the term as employed in philosophy.

[2] One can find the roots of such an attitude in at least some of the writers of the 19th and early 20th century. One might think here of Carl Wunderlich [51] or Ernst Romberg [36].

[3] "Ability to" here should be understood broadly to mean the ability to accomplish goals of individuals as opposed to physiological or psychological functions or dysfunctions which may, for example, underlie unacceptable pains. The ability to function in the second sense is also of importance in the elaboration of theories of disease; see, for example, [5].

[4] By clinical medicine I mean medicine focused on solving the problems of particular patients or of populations of patients. I would include here enterprises in preventive medicine, and exclude those of the basic sciences insofar as they are not concerned with applying medical knowledge to ameliorating the problems of particular patients or populations of patients. The line between applied and non-applied sciences is surely unclear, if not arbitrary. However, what is at stake is the concern that applied scientists have with the costs involved in the application or use of particular generalizations regarding data (here morbidity, mortality, and financial costs associated with over- and under-diagnosis). The issue is not that of simply applying knowledge expressed in universal statements to particulars, or of using scientific generalizations from other domains. Rather the issue, often overlooked, is that of fashioning descriptions of reality not simply for epistemic purposes, but for purposes of usefulness as well.

5 It is interesting to note that the earliest report of the use of artificial insemination is approximately two hundred years old ([34], p. 488). Kass's somatic problems would need to be extended to encompass how medicine has generally been practiced [26].

6 Clouser, Culver, and Gert have, in a very interesting article, offered a concept of malady as the genus to which medical problems such as disease, illness, dysfunction, handicap, injury, and sickness belong. "A person has a malady if and only if he or she has a condition, other than a rational belief or desire, such that he or she is suffering, or is at increased risk of suffering, an evil (death, pain, disability, loss of freedom or opportunity, or loss of pleasure) in the absence of a distinct sustaining cause" ([12], p. 36). The difficulty is in part that this definition is too restrictive. One is treating a clinical problem, for example, when one gives artificial insemination to a woman whose husband is sterile. A clinical problem can have a distinct sustaining cause. Moreover, what one will count as a distinct sustaining cause will depend upon one's interests. Is the presence of a phenylalanine-rich diet a distinct sustaining cause of the mental retardation of a child with PKU? A great number of genetic problems, as well as other maladies, require particular environmental background conditions. One might compare PKU with the inability of humans, versus some other mammals, to produce endogenous vitamin C. Does a universal genetic "defect" cease to be a disease or a defect when the trait is universal? For that matter, there are a number of maladies dependent upon distinct sustaining causes if scurvy is a malady. Moreover, what would individuals such as Clouser, Culver, and Gert say of very fair-skinned Swedes living in tropical Africa? Are they similar to individuals who have allergies to allergens present in certain environments, since in such an environment they would be exposed to dermatological problems, including a higher risk of squamous cell carcinoma? What will count as causes or background conditions will in part depend upon one's interests [24]. In short, one will not be able to cut as clearly as Clouser, Culver, and Gert would hope with their notion of a "distinct sustaining cause".

7 "Proceeding from a non-empty set of data D at a time-point t_1, which we will call D_{t_1}, the physician chooses from the infinite set A of all possible *actions* which one would consider, the particular sub-set which we will call A_1, and accomplishes it. The results of these actions carry the original set of data D_{t_1} over into the set of data D_{t_2}, at the time point t_2. Proceeding from D_{t_2} a second sub-set A_2 is chosen from A and rendered actual, etc. The finite sequence of this acquisition of data at time-points t_1, ..., t_n and the selection of actions

$$D_{t_1}, A_1, D_{t_2}, A_2, \ldots, D_{t_n}, A_n$$

is called the 'clinical process', where n is equal to or greater than 1. The fundamental problem of the clinical process is the question: is it possible to construct an effective procedure p which can be initiated at time-point t_1 concerning the data set D_{t_1} available concerning a patient where 1 is less than or equal to i, which is less than or equal to n, so that the optimal set of actions A_i can be selected unambiguously from the set of A of all possible actions, so that the person of the physician is exchangeable in this process? Or formulated in a different fashion: is there, for the set $D = D_{t_1}$, $D_{t_2}, \ldots, D_{t_n}$ of the data-sets at $t_1, \ldots, t_n$ concerning the patient and the potential set Pot (A) of the set A of all possible actions, a portrayal $v:D \to$ Pot (A), that will render the clinical process a rule-governed behavior and unambiguously provide the physician in all possible clinical situations with an optimal guide for his decisions?" ([37], p. 77.)

BIBLIOGRAPHY

[1] American Psychiatric Association: 1980, *Diagnostic and Statistical Manual of Mental Disorders (DSM III)*, 3rd ed., American Psychiatric Association, Washington, D.C.

[2] Bichat, Xavier: 1801, *Anatomie générale, appliquée à la physiologie et la médecine*, 4 vols. in 2, Brosson, Paris.

[3] Boorse, Christopher: 1975, 'On the Distinction Between Disease and Illness', *Philosophy and Public Affairs* 5 (Fall), 49–68.

[4] Boorse, Christopher: 1976, 'What a Theory of Mental Health Should Be', *Journal for the Theory of Social Behavior* 6, 61–84.

[5] Boorse, Christopher: 1976, 'Wright on Functions', *The Philosophical Review* 85 (Jan.), 76–86.

[6] Boorse, Christopher: 1977, 'Health as a Theoretical Concept', *Philosophy of Science* 44 (Dec.), 542–573.

[7] Broussais, F. J. V.: 1824, *Examen des doctrines médicales et des systèmes de nosologie*, 2 vols., Mequignon-Marvis, Paris.

[8] Broussais, F. J. V.: 1828, *De l'irritation et de la folie*, Delaunay, Paris; *On Irritation and Insanity*, McMorris Press, Columbia, South Carolina, 1831, Thomas Cooper (tr.).

[9] Burns, Chester R.: 1975, 'Diseases Versus Healths: Some Legacies in the Philosophies of Modern Medical Science', in H. T. Engelhardt, Jr. and Stuart F. Spicker (eds.), *Evaluation and Explanation in the Biomedical Sciences*, D. Reidel Publishing Company, Dordrecht, pp. 29–50.

[10] Canguilhem, G.: 1972, *Le normal et le pathologique*, Presses Universitaires de France, Paris.

[11] Caplan, A. L.: 1981, 'The "Unnaturalness" of Aging – A Sickness Unto Death?', in A. L. Caplan, H. T. Engelhardt, Jr. and J. J. McCartney (eds.), *Concepts of Health and Disease*, Addison-Wesley Publishing Co., Reading, Massachusetts, pp. 725–738.

[12] Clouser, K. D., C. M. Culver and E. Gert: 1981, 'Malady: A New Treatment of Disease', *The Hastings Center Report* 11 (June), 29–37.

[13] Cohen, H.: 1961, 'The Evolution of the Concept of Disease', in B. Lush (ed.), *Concepts of Medicine*, Pergamon Press, Oxford, pp. 159–169.

[14] Cullen, W.: 1769, *Synopsis nosologiae methodicae*, William Creech, Edinburgh.

[15] Engelhardt, H. T., Jr.: 1975, 'The Concepts of Health and Disease', in H. T. Engelhardt, Jr. and S. F. Spicker (eds.), *Evaluation and Explanation in the Biomedical Sciences*, D. Reidel Publishing Co., Dordrecht, pp. 125–142.

[16] Engelhardt, H. T., Jr.: 1976, 'Human Well-Being and Medicine: Some Basic Value-Judgements in the Biomedical Sciences', in H. T. Engelhardt, Jr. and D. Callahan (eds.), *Science, Ethics, and Medicine*, Institute of Society, Ethics and the Life Sciences, Hastings-on-Hudson, New York, pp. 120–139.

[17] Engelhardt, H. T., Jr.: 1980, 'Doctoring the Disease, Treating the Complaint, Helping the Patient: Some of the Works of Hygeia and Panacea', in H. T. Engelhardt, Jr. and D. Callahan (eds.), *Knowing and Valuing: The Search for Common Roots*, Institute of Society, Ethics and the Life Sciences, Hastings-on-Hudson, New York, pp. 225–249.

[18] Engelhardt, H. T., Jr.: 1981, 'Clinical Judgment', *Metamedicine* **2**, 301–317.

[19] Faber, K.: 1923, *Nosography in Modern Internal Medicine*, Paul F. Holber, New York.

[20] Foucault, M.: 1963, *Naissance de la Clinique*, Presses Universitaires de France, Paris.

[21] Gedye, J. L.: 1979, 'Stimulating Clinical Judgment: An Essay in Technological Psychology', in H. T. Engelhardt, Jr., S. F. Spicker and B. Towers (eds.), *Clinical Judgment: A Critical Appraisal*, D. Reidel Publishers, Dordrecht, pp. 93–114.

[22] Goosens, W. K.: 1980, 'Values, Health, and Medicine', *Philosophy of Science* **47**, 100–115.

[23] Grene M.: 1978, 'Individuals and Their Kinds: Aristotelian Foundations of Biology', in S. F. Spicker (ed.), *Organism, Medicine, and Metaphysics*, D. Reidel Publishers, pp. 121–136.

[24] Hart, H. L. and Honoré, A. M.: 1959, *Causation in the Law*, Oxford University Press, Oxford.

[25] *Hippocrates*: 1923, W. H. S. Jones (tr.), Harvard University Press, Cambridge, Massachusetts, Vol. 1.

[26] Kass, L.: 1975, 'Regarding the End of Medicine and the Pursuit of Health', *The Public Interest* **40** (Summer), 11–24.

[27] King, L. S.: 1954, 'What Is a Disease?', *Philosophy of Science* **21** (July), 193–203.

[28] Kistner, R. W.: 1973, 'The Menopause', *Clinical Obstetrics and Gynecology* **16** (Dec.), 107–129.

[29] Koch, R.: 1920, *Die ärztliche Diagnose*, J. F. Bergman, Wiesbaden.

[30] Linnaeus, C.: 1763, *Genera morborum, in auditorum usum*, Steinert, Upsaliae.

[31] Margolis, J.: 1976, 'The Concept of Disease', *The Journal of Medicine and Philosophy* **1**, 239–255.

[32] Niebyl, P.: 1971, 'Sennert, Van Helmont, and Medical Ontology', *Bulletin of the History of Medicine* **45** (March–April), 115–137.

[33] Pagel, W.: 1958, *Paracelsus*, S. Karger, Basel.

[34] [The] *Philosophical Transactions of the Royal Society of London*: 1809, Vol. XVIII, C. and R. Baldwin, London.

[35] Post, R. H.: 1962, 'Population Differences in Red and Green Color Vision Deficiency: A Review, and a Query on Selection Relaxation', *Eugenics Quarterly* **9**, 131–146.

[36] Romberg, E.: 1909, *Lehrbuch der Krankheiten des Herzens und der Blutgefässe*, 2nd ed., Ferdinand Enke, Stuttgart, p. 4.

[37] Sadegh-zadeh, K.: 1977, 'Grundlagenprobleme einer Theorie der klinischen Praxis', *Metamed* **1**, 76–102.

[38] Sadegh-zadeh, K.: 1977, 'Krankheitsbegriffe und nosologische Systeme, *Metamed* **1**, 4–41.

[39] Sauvages de la Croix, François Boissier de: 1763, *Nosologia methodica sistens morborum classes juxta Sydenhami mentem et botanicorum ordinem*, 5 vols., Fratrum de Tournes, Amsterdam.

[40] Sauvages de la Croix, François Boissier de: 1768, *Nosologia methodica sistens morborum classes juxta Sydenhami mentem et botanicorum ordinem*, 2 vols., Fratrum de Tournes, Amsterdam.·

[41] Sydenham, T.: 1676, *Observationes medicae circa morborum acutorum historiam et curationem*, G. Kettilby, London.

[42] Taylor, F. K.: 1979, *The Concepts of Illness, Disease and Morbus*, Cambridge University Press, Cambridge.

[43] Temkin, O.: 1963, 'The Scientific Approach to Disease: Specific Entity and Individual Sickness', in A. C. Crombie (ed.), *Scientific Change*, Basic Books, New York, pp. 629–647; reprinted in A. Caplan *et al.* (eds.), 1981, *Concepts of Health and Disease*, Addison-Wesley, Reading, Massachusetts, pp. 247–263.

[44] Trivers, R: 1974, 'Parent-Offspring Conflict', *American Zoologist* 14, 249–264.

[45] Virchow, R.: 1847, 'Über die Standpunkte in der wissenschaftlichen Medicin', *Archiv für pathologische Anatomie und Physiologie und für klinische Medicin* 1, 3–7, S. G. M. Engelhardt (tr.), in A. Caplan, H. T. Engelhardt, Jr. and J. J. McCartney (eds.), *Concepts of Health and Disease*, Addison-Wesley Publishers, Boston, Massachusetts, pp. 187–190.

[46] Virchow, R.: 1895, *Hundert Jahre allgemeiner Pathologie*, Verlag von August Hirschwald, Berlin, pp. 35–41, S.G.M. Engelhardt (tr.) in A Caplan, H. T. Engelhardt, Jr. and J. J. McCartney (eds.), *Concepts of Health and Disease*, Addison-Wesley Publishers, Boston, Massachusetts, pp. 190–195.

[47] von Wright, G. H.: 1963, *The Varieties of Goodness*, Humanities Press, New York.

[48] Williams, G. C.: 1966, *Adaptation and Natural Selection*, Princeton University Press, Princeton, New Jersey.

[49] Wilson, R. A., R. E. Brevetti and T. A. Wilson: 1963, 'Specific Procedures for the Elimination of the Menopause', *Western Journal of Surgery, Obstetrics and Gynecology* 71 (May–June), 110–121.

[50] Wulff, H.: 1981, *Rational Diagnosis and Treatment*, 2nd ed., Blackwell Scientific Publishers, Oxford.

[51] Wunderlich, C. A.: 1842, 'Einleitung', *Archiv für physiologische Heilkunde* 1, ix.

KAZEM SADEGH-ZADEH

COMMENTS ON ENGELHARDT'S 'CLINICAL PROBLEMS AND THE CONCEPT OF DISEASE'

From the many intriguing and important ideas Professor Engelhardt presents in his excellent paper, I shall choose to comment on only the following four: (1) ontological and anti-ontological views on disease; (2) value-ladenness of pathology and nosology; (3) the concept of clinical problems as a substitute for the traditional concept of disease; (4) clinical medicine as an applied science.

I

Professor Engelhardt discusses first the problems arising in the ontological debates about the nature of 'disease'. He then goes on to conclude that one might also construe such debates as disputes regarding the most useful ways in which to approach medical reality, where usefulness is being assessed not only in terms of the capacity to develop further explanations of disease, but useful treatments for patients as well (p. 33). Though this is a very interesting idea and might contribute to a philosophically enlightening reconstruction of the recent history of medicine, I do not believe that it is a sufficient basis for understanding why so many physicians and philosophers of medicine dispute about the reality and unreality of disease. I think that these disputes are mainly due to two circumstances: (a) to a conceptual confusion reflecting the lack of an explicitly formulated and generally accepted concept of disease in medicine; and (b) to the neglect of the ontological problems associated with the notion of 'existence'.[1] Despite these semantically and ontologically interesting philosophical themes, I completely agree with Professor Engelhardt's implicit opinion that all versions of disease ontology are practically irrelevant to medicine in general and clinical practice in particular.

II

Professor Engelhardt's insistence on the value-ladenness of pathology and nosology deserves much interest even though there are many people who cannot understand why a disease should be viewed as a value-laden entity.

43

L. Nordenfelt and B.I.B. Lindahl (eds.), Health, Disease, and Causal Explanations in Medicine, 43–45.

I would like to propose a mode of reasoning which reduces the necessity for lengthy arguments with the adherents of the value-neutral position. The following premise is a true empirical statement about the behavior of the members of every existing population X:

> An entity which is not value-laden in a population X is not viewed as a disease in this population.

This true premise, which even Christopher Boorse cannot reject, yields by contraposition:

> An entity which is viewed as a disease in a population X is value-laden in this population.

This is exactly Professor Engelhardt's thesis that independent of its particular definition in particular societies, every concept of disease is value-laden.[2]

III

Professor Engelhardt proposes to substitute concepts of clinical problems for the traditional concept of disease (p. 33). This proposal is a very interesting one and might contribute to a beneficial reduction of misleading ideas on the one hand, and to an efficient structuring of medical textbooks and education on the other. For example, I can imagine a medical textbook on internal medicine whose chapters are devoted to the description, analysis and treatment of patients' complaints instead of the presentation of obscure morbi. However, I am afraid that after one hundred years or so a concept of disease would arise once again and one would begin to construct new diseases and to re-substitute them for clinical problems. In other words, I believe that the invention of diseases as 'causes' of clinical problems has been epistemologically unavoidable in the history of medicine. Let me substantiate this belief with Professor Engelhardt's own excellent words. In the history of medicine, "One will wish to isolate from a complex of causally relevant factors those that ought to be held therapeutically accountable" (p. 35). And exactly these supposedly causally relevant factors whose therapeutic control is held to be practically relevant to an efficient management of Professor Engelhardt's clinical problems, are traditionally viewed as diseases.

IV

Let me conclude with a few words about Professor Engelhardt's understanding of clinical medicine as an applied science (p. 35). I think it is beneficial to refine this widespread interpretation in the following way.

The term 'applied science' is vague and has at least two meanings: firstly, a science which applies knowledge that is offered by other disciplines like physics, chemistry, etc.; secondly, a science whose nature consists in the application of knowledge to particulars. In either case, the characterization of clinical medicine as an applied science is misleading. Clinical medicine consists of two parts: (a) clinical research, and (b) clinical practice.

Ad (a): Clinical research is not an applied science in the sense that it employs scientific results from other disciplines. It is something more than mere applied science. It aims at discovering efficient rules of practice and replacing available rules of practice by more efficient ones. Thus, it is a science of efficient practice, or, a practical science in the sense of Aristotle.

Ad (b): Clinical practice, however, is not a science at all, and consequently not an applied science, though scientific knowledge is – or may be – applied there. Clinical practice is a field of more or less scientifically-based moral decisions and actions. Thus, it is a particular field of practical ethics and politics.[3]

University of Münster,
Federal Republic of Germany

NOTES

[1] Cf. my 'Krankheitsbegriffe und nosologische Systeme', *Metamed* **1** (1977), 4–41.
[2] In the present context, an elaborate explication of the notion of value-ladenness is not necessary. It suffices to define it as 'viewed as beneficial or risky'. For a more comprehensive account, see my 'Normative Systems and Medical Metaethics. Part 1: Value Kinematics, Health, and Disease', *Metamedicine* **2** (1981), 75–119.
[3] See my concept of 'differential praxiognosis' in my 'Foundations of Clinical Praxiology. Part 1: The Relativity of Medical Diagnosis', *Metamedicine* **2** (1981), 183–196. A more elaborate discussion of these two points may be found in my 'Medicine as Ethics and Constructive Utopia. Part 1' (in German), *Medizin, Ethik & Philosophie* **1** (1983), 1–18.

RALPH GRÄSBECK

HEALTH AND DISEASE FROM THE POINT OF VIEW OF THE CLINICAL LABORATORY *

1. THE EVOLUTION OF THE REFERENCE VALUE CONCEPT

I became involved in the philosophical problems of medicine by introducing together with Saris in 1969, rather boldly and unsuspectingly, the concept of reference values [20]. As arrangers of a congress we brought up the topic because we thought it might be interesting. The mental process leading up to this concept can be traced back to about 10 years earlier when several facts attracted my attention: I had been taught that Nature, including the results of laboratory tests, was distributed in a Gaussian manner, but in a study on serum vitamin B_{12} we found that this was not so, which puzzled me [19]. Second, when returning to clinical work from experimental research I noticed that the controls were unsatisfactory. In experimental work you meticulously arrange that the controls are very similar to the objects of the actual experiment. Diseases and patients may be regarded as experiments made by Nature, but the controls, and the data derived from them, i.e., the normal values, were used without paying attention to whether they were comparable to the patients and their data. The normal values were usually derived from young, ambulant persons, frequently hospital personnel, whereas the patients tended to be old and bedridden. Third, I realised that the word normal was ambiguous, meaning, e.g., frequent, non-pathological and Gaussian [18] and others agreed ([24], [29]). The term 'reference values' represents a wider concept than normal values covering data from both healthy, unselected and diseased individuals or groups of individuals, "negative and positive clinical controls". There is also the rigid requirement that both the individual, the conditions of the collection and analysis of the specimen and the mathematical treatment of the data be stated in detail and made available to the user ([6], [22]).[1] The new term and the corresponding philosophy have been further developed as a collective effort and are now accepted and recommended by the International Federation of Clinical Chemistry [22]. The first recommendation demands that the "state of health" of the reference individuals be described. In order to apply this recommendation and to categorise the control[2] individuals and the values derived from them it is necessary to understand and define the concepts of health and disease.

47

L. Nordenfelt and B.I.B. Lindahl (eds.), Health, Disease, and Causal Explanations in Medicine, 47–60.
© 1984 by D. Reidel Publishing Company.

2. WHY ARE LABORATORY TESTS ORDERED?

The general reasons for ordering laboratory tests are rarely discussed. In most medical institutions the tests are ordered not to decide whether a person is ill or healthy, but to monitor the development of a disease and the effects of therapy. Unless the doctor works in the armed forces or a prison, he can usually rely on the patient's statement that he is ill. A reason for ordering laboratory tests is therefore not to distinguish between health and disease but one disease from another.[3] In contrast to the situation in hospitals, distinguishing between healthy and sick individuals is of central importance for health insurance and in epidemiology and preventive medicine. The latter branch of medicine is rapidly expanding. Surveillance of pregnancy, prevention of occupational disease, screening for tuberculosis may be offered as examples. The use of laboratory tests for such purposes is increasing, but few tests are really useful. All laboratory tests tend to give false positive and false negative results and Bayes' theorem tells us that when the prevalence of a disease is low even the use of a fairly specific test is expensive and may cause trouble ([9], [34]), e.g., psychological problems. None of us would enjoy the experience of being told he has a positive test for cancer or syphilis and later "so sorry, the test was wrong".

3. HEALTH

3.1. Importance of Defining Health: The Goal

Even so, reference values from persons who are healthy are of great importance. One way of deciding that the patient is getting better and that the treatment is efficient is to observe that his laboratory tests give results similar to those of healthy persons. In laboratory medicine (as in medicine in general) health represents the goal and health-associated reference values are goal values ([15], [16]).

However, what this goal really is, is a complicated question. The well-known definition of health in the WHO constitution is widely recognized as being unrealistic and is not even taken seriously by WHO itself since it professes to achieve "health for everybody in the year 2000" which is clearly impossible, since taken literally, it means that nobody shall die that year. At the bedside the immediate goals of the doctor are usually obvious; he is concerned to cure the person of his acute distress. For instance, if the patient suffers from pneumonia or appendicitis the doctor attempts to eliminate this disease. When it comes to chronic conditions such as arthrosis (worn joints), atherosclerosis and diabetes, which cannot be cured, the doctor attempts to alleviate the condition, preferably to restore the capacity of the patient to work, lead an ordinary life, to be able to get along at home or simply to prevent him from dying. The table below presents two realistic and goal-oriented definitions of health that I have suggested. The health-related reference values have to be collected being aware of the current goal.

TABLE

Definition of 'health' (to replace the WHO definition)

---

1. Health is characterised by a minimum of subjective feelings and objective signs of disease, assessed in relation to the social situation of the subject and the purpose of the medical activity, and is in the absolute sense an unattainable ideal state.

2. Health is an abstraction representing the chief goal of medicine.

Definition 1 expresses current views among biomedical scientists. Definition 2 is aphorism-like and attempts to express the current use of the word "health" by politicians and administrators. The expression "chief goal" (healing, maintenance of well-being) is used because medicine may have other goals such as research prompted entirely by curiosity and biological warfare.

From Gräsbeck [16].

3.2. Partial or Total Function, Subjective and Objective Health

Health is often assessed taking into account the total function of the body ([3], [8]). According to this philosophy a person who lacks one kidney or has a therapeutically well-controlled diabetes is well. This holistic (overall) approach is popular among administrators, social workers and those concerned with nursing, but is rarely applicable in laboratory medicine which looks at partial functions, e.g., how an organ or one metabolic system functions. Measurement of working capacity, physical performance, is perhaps an exception [16]. Health can also be subjective or objective, or both. If the 'diagnosis' of health is to be based on subjective feelings we must note that the will of the subject plays a decisive role. Some persons don't care about severe symptoms and insist on working, others use every excuse to absent themselves from work. Some psychiatric conditions and brain diseases are characterized by euphoria, an increased feeling of well-being ([15],

[16]). Though subjectively healthy people are often used as sources of "normal values", the objective character of laboratory medicine discourages us from continuing with this practice, and objective observations must be included among the health criteria.

3.3. Age, Ethical Issues

The result of many laboratory tests, e.g., the red cell count and the haemoglobin concentration in blood vary with age. Two periods are characterized by great variations: infancy through puberty to early adulthood and the period after the menopause in women when many values, e.g., alkaline phosphatase in serum, tend to approach those found in men [17]. In some cases it is therefore customary to compare observed values with "normal" values from persons of the same age; this is common practice in the case of children and the results of sex hormone assays. Because of improved analytical accuracy and specificity, variations with age have recently been revealed for other components. In interpreting values for adults, and especially the aged, the problem of which age group the corresponding health-related reference values should be derived from arises. It has been suggested that certain components should be followed during a person's entire lifetime and that the reference or goal should be the values observed during his "most healthy state" [30]. This period occurs during young adulthood. However, let us pursue the matter further. Clinical laboratories traditionally assay substances in blood and urine, but in a strict sense these fluids do not represent the subject himself. We, i.e., our cells, live in a tissue culture fluid, the blood plasma, nowadays considered to correspond roughly to the sea where life once evolved. Claude Bernard coined the term 'milieu intérieur' for it. However, clinical chemistry has begun to analyse tissues and cells rather than the pond where we swim. Today it is already possible to stick a thin needle into almost any organ, including the heart and the brain, and to draw and analyse an extremely small specimen. The aging process takes place in the tissues and it is likely that in contrast to the slight changes occurring in the milieu intérieur more dramatic alterations will be observed in some types of cells. If we now begin to adjust the composition of the tissues with therapy to correspond to that in healthy young individuals it means that we are trying to rejuvenate the old and perhaps keep them alive forever. Whether we should accept this goal or not is an ethical question. Personally, I feel that for ethical and economical reasons we should go on comparing laboratory results (observed values) with reference values from roughly the same age category ([15], [16]).

3.4. Longitudinal Assessment of Health

But, what criterion of health should we use for the aged who almost all suffer from some kind of ailment? I have suggested longitudinal assessment of health [10]. Tests are performed on persons who are relatively healthy and then the values, or possibly the specimens, are saved for, say, 10 years. Then we use the values for the persons who have not died or become seriously ill.

Similar goal-orientated survival reference values could be used for premature babies, because the state of prematurity is not "normal". But again ethical problems crop up: these babies (and also healthy ones) cannot give informed consent to taking specimens for determination of reference values [21] and their blood volume is so small that drawing of blood may cause anaemia [5].

3.5. Risk and Ideal Values

Research, especially on cardiovascular disease, has revealed that certain values (e.g., serum lipid concentrations and blood pressure readings) statistically predispose to disease such as atherosclerosis and its complications including cerebral stroke and myocardial infarction. In this case as well as generally there is no dichotomy between the values associated with high and low risk, but there is a smooth change in the risk level as the values of the laboratory tests increase [2]. However, one can arbitrarily decide upon a risk limit and draw a reference limit below (or sometimes above) which the values are statistically associated with a low or moderate risk. Such values are even more goal-orientated than the ordinary health-associated reference values. As to ethics, preventing cardiovascular accidents is a generally accepted goal, though representatives of the third world have pointed out that prevention of malnutrition and tropical diseases would be more important.

3.6. Biochemical Individuality

If you compare the concentrations of components in blood at different times you usually observe that the intra-individual variation is much smaller than the inter-individual variation ([30], [32]). The difference between individuals is sometimes amazing. These differences are caused by "biochemical individuality" [31]. Some of the individuality is based on true qualitative chemical differences. For instance, the surfaces of our cells contain chemical structures such as blood group and HLA (transplantation) antigens, which differ in different individuals. The different antigens are clearly associated with different frequencies of certain diseases. For instance, some HLA-types predispose·to intestinal malabsorption and rheumatoid disease. Metabolism can be described as follows. (See Figure 1.) Substance *i* is converted by

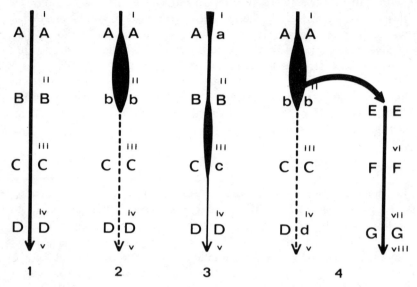

Fig. 1. Metabolism and biochemical individuality. The pairs of letters (e.g., AA, Dd, bb) denote gene pairs and the corresponding enzymes coded by them. Capital letters denote "good" genes, small letters "bad" or absent ones. In the case of enzymes, two capital letters mean "usual or optimal activity", capital plus small roughly "half activity" and two small "no or poor activity". The metabolites are indicated by Roman numerals, their amounts by the width of the arrows. 1 – Optimum situation. 2 – Congenital error of metabolism with accumulation of precursors and lack of products. Enzyme B is absent or inactive. 3 – Heterozygotism with hypothetical slight increase of precursors (especially in load situations) and relative lack of products. Enzyme C is present in roughly half the optimum concentration. 4 – Compensation by "side track" metabolism. This is a simplified scheme. There are mechanisms that switch the genes on and off. However, these mechanisms are genetically controlled by "good" and "bad" genes, too.

enzyme *A* into substance *ii*, *ii* in turn by enzyme *B* into substance *iii*, etc. The enzymes are proteins and their structure is coded in the corresponding genes. With some exceptions, we have two genes for every protein, one from our father and the other from our mother. If both genes are abnormal or lacking, the corresponding enzyme is abnormal or lacking, which usually means that the precursor (e.g., substance *ii*) piles up and the products in the chain after the abnormal or lacking enzyme (e.g., *B*) are not produced at all or in insufficient quantities. This causes a congenital error of metabolism. Numerous such diseases have been described. The situation where only one of the two genes is abnormal or lacking, heterozygotism, must therefore be extremely common. In fact, it is likely that we are all heterozygotes in respect to one disease or another. Nowadays clinical chemistry is often capable of demonstrating heterozygotism because the enzyme or protein in question has a low concentration. Biochemical individuality is probably partly due to this quantitative factor.

3.7. Nobody is Completely Healthy

I suspect that such heterozygotism may not be completely harmless though as with defects of other kinds, it may be compensated [25]. It acts throughout our lifetime and may cause a gradual accumulation of precursors and a relative lack of products and perhaps ultimately result in a chronic disease such as atherosclerosis. It may reduce our tolerance to external factors such as foodstuffs and drugs and predestine us to the diseases we shall ultimately succumb to.

For this reason and considering the aging process (which, incidentally, could be regarded as a disease with a fatal outcome) I maintain that nobody is completely healthy, some unhealthiness is present in everybody as "entropy in chemical systems" [10]. Health, like temperature, may be higher or lower [1], but as long as we live, we possess some health. Health can only be diagnosed by exclusion (i.e., it is a privative concept) [33], but if we really do a thorough job, we will find at least a minor defect. I also conclude that a person may be healthy in some respects and ill in other respects. He may, e.g., be a perfect, healthy control for rheumatic disease or glucose metabolism but unacceptable for cardiovascular disease or cholesterol metabolism.

3.8. Practical Problems

Many practical problems arise when health has to be "diagnosed". What is usually done is that you ask persons in your vicinity [7], colleagues, laboratory technologists, etc.: "Are you all right?" If they say yes, you tell them to come back the next morning without having eaten and then you collect some blood. This means that you examine people of working age and use working capacity and subjective health as selection criteria. However, many of these people are not really well. People take a lot of drugs, for instance aspirin against menstrual disorders or hangover, undergo permanent treatment for mild rheumatoid arthritis, hypertension etc., or take tranquilizers. Everyday life often involves intake of nicotine, caffeine and ethanol, which are pharmacological agents, and ethanol in particular influences several components in serum strongly and for several days. Contraceptive steroids act much like pregnancy and profoundly influence laboratory results ([27], [28]).

Traditionally even a very superficial health check-up includes determination of glucose and protein in the urine to exclude diabetes and kidney disease. The question arises, is it philosophically acceptable to include such tests in selecting healthy controls for laboratory tests? The Scandinavian Society of Clinical Chemistry and Clinical Physiology has a Reference Values Committee that has discussed the problem at length and produced a recommendation [26]. In this recommendation we use certain laboratory tests including those mentioned as criteria of health. Diabetes and kidney disease have profound influences on several components in blood and urine and common sense would direct us not to include such patients among healthy controls. Also, two laboratory tests may be just as independent of each other as, say, an anthropometric measure such as hair colour and a laboratory test such as erythrocyte sedimentation rate.

Our committee also has practical experience of collecting reference values from a general population. In Kristianstad in Southern Sweden a long-term health survey project has been pursued and recently, the health criteria set up by our committee [26] were applied. The criteria were based on the well-known WHO list of diagnoses and were of the type that a person who had suffered from myocardial infarction or severe tuberculosis or recently had flu or taken drugs was eliminated as a healthy control. It turned out that the fraction of the population remaining after the elimination was so small that it was almost useless [4]. It is evident that extensive studies are needed to elucidate which components are affected in which diseases and which are not. As stated above, the subjects may then serve as healthy controls for some purposes and not be acceptable for others.

4. DISEASE

4.1. Relativity and Subjectivity

Let us now turn to disease or unhealth, the antithesis of health. Again it can be subjective or objective, affect different metabolic systems, etc. It is also relative to the patient's age, profession and geographical location.

It is important to have available knowledge about typical values associated with certain diseases. For instance, we need to know the characteristic thyroid hormone values in persons whose thyroid glands function in an acceptable way, too little and too much (euthyroidism, hypothyroidism and hyperthyroidism). However, no matter which kind of thyroid test you use, the histograms or the curves describing the distribution of the values in these typical populations, always overlap [23]. Second, the diagnosis of for instance hyperthyroidism, involves a lot of subjectivity: The patient has complaints, e.g., his heart beats too fast, he has diarrhoea, feels hot, loses weight, and the doctor thinks he has sweaty hands, his eyes protrude, etc. Then the doctor orders laboratory tests, most of which are roughly equivalent to determining thyroid hormones. Having obtained the results he decides that the patient is hyperthyroid; this is based on a subjective evaluation of all findings. Subsequently, the condition is treated and the patient perhaps (fortunately frequently) feels better and also the doctor thinks his state has improved. Ergo, the patient suffered from hyperthyroidism. However, closer examination of the decision process reveals that there is circular reasoning and plenty of subjectivity in both the patient and the doctor. Perhaps the improvement was due to psychological influence, placebo effect? Finally, both the healthy and hyperthyroid reference individuals were chosen and categorized in a subjective fashion. Laboratory medicine is less objective than we tend to believe.

4.2. Taxonomy, Constant Breakdown of Entities

The classification of diseases has deep roots stretching down to the early history of medicine. Some are just learned words for symptoms (e.g., diarrhoea), others for macroscopic or microscopic findings in tissues (e.g., carcinoma), some are based on the causative agent (e.g., tuberculosis), etc. [34]. Diagnoses are constantly breaking down into new entities [13]. An example: Much of my own research has been concerned with pernicious anaemia. This condition was originally described as being characterized by typical changes in the peripheral blood and the blood-forming organ, the bone marrow. Research performed after the Second World War has shown that most of these cases are due to lack of vitamin B_{12}, but morphologically similar conditions can be produced by lack of another vitamin, folic acid. The most common cause of vitamin B_{12} deficiency is lack in gastric juice of a substance called Castle's intrinsic factor, which stimulates the absorption of the vitamin in the intestine. The usual cause of the lack of the intrinsic factor is degeneration in the mucosa of the stomach, which is in turn usually due to an autoimmune process in the mucosa in which hereditary factors play a role. This complicated constellation leading to vitamin B_{12} deficiency is roughly what pernicious anaemia means *today*. However, the intrinsic factor may also be the lack congenitally and then the stomach mucosa looks "normal". Finally, this substance may be misconstructed due to abnormal genes. Thus this originally haematological disease has broken down into numerous sub-diseases, which have little to do with the blood.

The situation is typical. There is a disease called pseudopseudohypoparathyroidism. Such a monstrous name tells you at once that the diagnosis of hypoparathyroidism has been subdivided into more and more entities. Therefore, the typical disease values of laboratory tests have to be constantly revised.

4.3. Levels of Diagnosis

In laboratory medicine it is practical to distinguish between different levels of diagnosis. First we try to achieve a rough classification, for instance search for anaemia by determining the red cell and haemoglobin content of whole blood or for kidney disease by assaying protein in the urine. Such tests are called screening tests and we strive for simplicity, cheapness and speed. Then we proceed to middle-level tests such as determining the size and shape of the red cells or assaying serum creatinine or urea. Then, if these tests indicate abnormality we proceed and try to get to what is today regarded as the final diagnosis, for instance, pernicious anaemia or autoimmune nephritis. However, an ambitious doctor or a university department usually tries to go deeper. For instance, we could try to elucidate what was the basic mechanism producing a case of pernicious anaemia, perhaps with the aim of publishing the case. The deeper you dive into the diagnosis, the more your activity becomes research. For instance, you may ask "why autoimmunity?" Nobody knows today. At some stage the authorities may interfere and tell you that the tests you are doing are a luxury and too expensive.

Also the "diagnosis" of health may have different levels ([12], [14]) depending on whether we accept subjective health alone, examine the subject superficially or thoroughly. However, complete certainty can never be reached. At some point the diagnostic procedures including the laboratory tests become so invasive and/or expensive that they are unacceptable, and they would not give complete certainty, anyway.

4.4. Alternative Categorization

The same diseases can be categorized in different ways. I offer tuberculosis as an example. In some, but not all, infected individuals the tubercle bacillus *Mycobacterium tuberculosis* causes a chronic inflammation of the lungs; the old name of this condition is phthisis. Though numerous persons are still exposed to this bacillus this disease is no longer very common in the developed countries. Its gradual disappearance is related to improved socio-economic conditions, BCG-vaccination, X-raying of the population, etc. However, certain population groups have an increased risk, especially old people, alcoholics and persons who receive immunosuppressive treatment, people living in misery and who starve, etc. The common denominator for these sensitive subjects could be called poor immune response, immunological insufficiency [13]. The same persons might just as well have got another infection, e.g., a staphylococcal hospital infection, hepatitis, leptospirosis, etc. In that case, why don't we call the condition immunological insufficiency? The main reasons are that the therapy of tuberculosis requires special drugs, traditionally tuberculotics are treated in special hospitals, and special legislation was passed at times when the disease was a great problem. The disease could also be regarded as just another kind of infectious lung disease, and that is actually the trend today. The diagnosis of pulmonary tuberculosis is therefore a pragmatic choice among several possibilities. From the point of view of the laboratory the diagnosis is very suitable because it is relatively easy to detect the microorganism responsible. To measure immunological insufficiency is still almost impossible, but may not remain for much longer.

4.5. To Everyone His Own Disease

A disease is not a simple entity such as an animal or a plant species. It is the result of endogeneous and exogenous factors, the reaction of an individual to one or several causative factors [34]. My thesis is that in fact everybody has his own disease and that no two persons (except perhaps identical twins) really have the same disease. The categorization of diseases into diagnoses is based on practical considerations: We need standard patterns of dealing with our patients to put them through an established machinery of investigations, treatment and perhaps social and legislative arrangements. Laboratory medicine has to help the clinician in this classification work. However, laboratory medicine, especially the purely scientific, investigative part of it, constantly influences what the clinicians conceive as disease entities.

Actually, it does this so strongly that in a generation or two the taxonomy of the diseases will be very different from that utilized today.

5. SCIENTIFIC MEDICINE

Practical clinical medicine is an old art, based on a mixture of tradition and ancient beliefs, long practical experience, theoretical considerations and hard experimental facts. That much of it is still an art rather than a science is illustrated by great regional differences in its practice. Back in Finland we are quick to puncture the eardrum in patients with acute otitis media (inflammation of the ear); here in Sweden they prefer just to give antibiotics. In the U.S.A. there is lot of fuss about sterility in connection with a lumbar puncture (drawing spinal fluid), here we do it casually like puncturing a vein. However, the scientific way of working is steadily seeping upwards from the experimental biomedical sciences into clinical medicine. Being closest to the basic sciences, clinical laboratory medicine is the most objective and scientific part of "the art of healing". Strict definitions, statistical ways of thinking and appropriate controls are therefore being quickly adopted by the staffs of clinical laboratories. This attitude is also reflected in the way we conceive of health and disease.

Minerva Institute for Medical Research,
Helsinki, Finland

NOTES

* The author's research has been supported by the Sigrid Jusélius Foundation and the Nordic Clinical Chemistry Project (NORDKEM) founded by the Nordic Council to promote the quality of the performance in hospital laboratories in the Nordic countries.
[1] Publications [7], [10], [11], [12], [13], [14], [15], [16], [17], [18], [19], [20], [22], [26], represent different stages leading up to the author's present views on the concepts of health, disease and reference values.
[2] If I were to suggest a new term (and philosophy) today to replace the "normal value(s)", I would probably choose the term "control value(s)".
[3] Reference values from typical patient populations are therefore desirable.

BIBLIOGRAPHY

[1] Alström, T.: 1981, 'Evolution and Nomenclature of the Reference Value Concept', pp. 3–13 in [17].

[2]   Aromaa, A. and Maatela, J.: 1981, 'A Search for Optimum Values: Prognostic Evaluation of Reference Values', pp. 145–165 in [17].

[3]   Behr, W. and Herrman, U.: 1976, *Probleme der theoretischen Medizin*, Verlag Volk und Gesundheit, Berlin, GDR.

[4]   Berg, B., Nilsson, J.-E., Solberg, H. E. and Tryding, N.: 1981, 'Practical Experience in the Selection and Preparation of Reference Individuals: Empirical Testing of the Provisional Scandinavian Recommendations', pp. 55–64 in [17].

[5]   Blumenfeld, T. A.: 1981, 'An American Prospective of Pediatric Blood Specimen Collection', pp. 85–96 in [17].

[6]   Dybkaer, R.: 1972, 'Concepts and Nomenclature in Theory of Reference Values', *Scandinavian Journal of Clinical and Laboratory Investigation* 29, Suppl. 126, 19.1.

[7]   Dybkaer, R. and Gräsbeck, R.: 1973, 'Theory of Reference Values', *Scandinavian Journal of Clinical and Laboratory Investigation* 32, 1–7.

[8]   Eriksson, K.: 1979, 'Semantiska och kulturella aspekter på hälsobegreppet' ('Semantic and Cultural Aspects of the Concept of Health') (in Swedish), *Finska Läkaresällskapets Handlingar* 123, 74–81.

[9]   Galen, R. S. and Gambino, S. R.: 1975, *Beyond Normality: The Predictive Value and Efficiency of Medical Diagnoses*, Wiley, New York.

[10] Gräsbeck, R.: 1972, 'Types of Reference Groups', *Scandinavian Journal of Clinical and Laboratory Investigation* 29, Suppl. 126, 19.2.

[11] Gräsbeck, R. 1977, 'Reference Value Philosophy', *IFCC Newsletter*, Fasc. 16, 4–6.

[12] Gräsbeck, R.: 1977, 'Normalvärden, referensvärden och hälsa' ('Normal Values, Reference Values and Health') (in Swedish), *Nordisk Medicin* 92, 142–144.

[13] Gräsbeck, R.: 1978, 'Terminology and Biological Aspects of Reference Values', in E. S. Benson and M. Rubin (eds.), *Logic and Economics of Clinical Laboratory Use*, Elsevier, New York, pp. 79–90.

[14] Gräsbeck, R.: 1979, 'Hälsa och referensvärden' ('Health and Reference Values') (in Swedish), *Finska Läkaresällskapets Handlingar* 123, 65–73.

[15] Gräsbeck, R.: 1981, 'Referenzwerte und Gesundheit', *Klinische Chemie Mitteilungen, Deutsche Gesellschaft für Klinische Medizin* 12, 2–8.

[16] Gräsbeck, R.: 1981, 'Health as Seen from the Laboratory', pp. 17–24 in [20].

[17] Gräsbeck, R. and Alström, T. (eds.): 1981, *Reference Values in Laboratory Medicine. The Current State of the Art*, Wiley, Chichester, U.K.

[18] Gräsbeck, R. and Fellman, J.: 1968, 'Normal Values and Statistics', *Scandinavian Journal of Clinical and Laboratory Investigation* 21, 193–195.

[19] Gräsbeck, R., Nyberg, W., Saarni, M. and von Bonsdorff, B.: 1962, 'Lognormal Distribution of Serum Vitamin B_{12} Levels and Dependence of Blood Values on the B_{12} Level in a Large Population Heavily Infected with Diphyllobothrium Latum', *Journal of Laboratory and Clinical Medicine* 59, 419–429.

[20] Gräsbeck, R. and Saris, N.-E.: 1969, 'Establishment and Use of Normal Values', *Scandinavian Journal of Clinical and Laboratory Investigation* 26, Suppl. 110, 62.

[21] Hicks, J. M., Hammond, K. and Boeckx, R. L.: 1981, 'Pediatric Reference Values', pp. 297–309 in [17].

[22] International Federation of Clinical Chemistry (IFCC), Expert Panel on the Theory of Reference Values (Gräsbeck, R., Siest, G., Wilding, P., Williams, G. Z.

and Whitehead, T. P.): 1978, 'Provisional Recommendation on the Theory of Reference Values', *Clinica Chimica Acta* 87, 459f.–465f.

[23] Lamberg, B.-A., Heinonen, O. P., Liewendahl, K., Kvist, G., Viherkoski, M., Aro, A., Laitinen, O. and Knekt, P.: 1970, 'Laboratory Tests on Thyroid Function in Hyperthyroidism II. A Statistical Evaluation of 13 Different Variables', *Acta Endocrinologica*, Suppl. 146, 23–35.

[24] Murphy, E. A.: 1972, 'The Normal and the Perils of the Sylleptic Argument', *Perspectives in Biology and Medicine* 15, 566–582.

[25] Nordenfelt, L.: 1981, 'Disease and Abnormality – A Conceptual Study', pp. 25–32 in [17].

[26] Scandinavian Society for Clinical Chemistry and Clinical Physiology, Committee on Reference Values (Alström, T., Gräsbeck, R., Hjelm, M. and Skandsen, S.): 1975, 'Recommendations Concerning the Collection of Reference Values in Clinical Chemistry and Activity Report', *Scandinavian Journal of Clinical and Laboratory Investigation* 35, Suppl. 144, 1–44.

[27] Siest, G.: 1981, 'Strategy for the Establishment of Healthy Population Reference Values', pp. 45–53 in [17].

[28] Statland, B. E. and Winkel, P.: 1981, 'Selected Pre-analytical Sources of Variation', pp. 127–137 in [17].

[29] Vácha, J.: 1978, 'Biology and the Problem of Normality', *Scientia* 113, 823–846.

[30] Williams, G. Z.: 1972, 'The Nature and Use of Individual Person's Reference Limits in Clinical Chemistry', *Scandinavian Journal of Clinical and Laboratory Investigation* 29, Suppl. 126, 19.3.

[31] Williams, R. L.: 1956, *Biochemical Individuality. The Basis for the Genetotrophic Concept*, Wiley, New York.

[32] Winkel, P.: 1981, 'The Use of the Subject as His Own Referent', 65–78 in [17].

[33] von Wright, G. H.: 1963, *The Varieties of Goodness*, 2nd ed., Routledge and Kegan Paul, London.

[34] Wulff, H. R.: 1981, *Rational Diagnosis and Treatment*, 2nd ed., Blackwell, Oxford.

SECTION II

ON DEFINITION AND CLASSIFICATION
IN MEDICINE

UFFE JUUL JENSEN

A CRITIQUE OF ESSENTIALISM IN MEDICINE

1. THE VOCABULARY OF ESSENTIALISM

(1) Modern clinical theory has inherited a philosophical vocabulary shared in the past by empiricists and rationalists. 'Necessary' and 'sufficient' are the key terms of this vocabulary. Use of these terms is unavoidable in two crucial contexts: when specifying conditions for referring in a clear and exact way to entities in the world, and when specifying the causal conditions for the occurrence of entities in the world.

Terms referring to entities have to be defined by specifying a conjunction of characteristics, each of which is necessary, and which together are sufficient for the use of the term. (The main thesis of traditional essentialism.)

Causal explanations must contain either specification of sufficient *or* of necessary conditions for the occurrence to be explained. Though the analysis of causality in terms of conditions has come under heavy attack, all kinds of philosophical rescue attempts have been tried, most recently by Mackie with his sophisticated INUS-conditions [7].

(2) Recent clinical epidemiological literature retains that vocabulary even though it is widely recognized that diseases escape definition and even though the old ideal of causal explanation in medicine (genetic conditions as *sufficient* conditions of phenylketonuria) is recognized as false ([1], p. 36).

(3) But why all these manoeuvres to retain the old vocabulary? Why this display of theoretical fidelity when it is not for practical reasons?

The traditional vocabulary has been regarded as the ideal tool for mirroring the world in which we live — for the clinician the ideal tool for representing the world of disease entities and the causal chains behind them. That ontological picture (sharply separated disease entities with specific causes) is, however, not independent of the vocabulary. It is *implied* by the vocabulary itself.

The essentialist requirements with respect to definitions (that a definition states a conjunction of characteristics each of which is necessary and together sufficient for the use of the term) and the requirements with respect to causal explanations (that they, ideally, specify sufficient condition of occurrences) *imply* the ontological picture of a world of fixed, static building blocks glued together by a causal cement.

63

L. Nordenfelt and B.I.B. Lindahl (eds.), Health, Disease, and Causal Explanations in Medicine, 63–73.
© 1984 by D. Reidel Publishing Company.

(4) Philosophical vocabularies die hard. But what explains the vitality of the essentialist vocabulary described above? The vocabulary is the child of a static (pre-Darwinian) conception of nature. A conception embodied in artistic representation of an orderly, hierarchical world built up with sharply separated departments, a conception which has constituted the basis of philosophical thinking since antiquity.

It would not be fair to blame Aristotle, Linnaeus, Sydenham and the traditional taxonomists of medicine for having a pre-Darwinian view of nature. But why is it, 100 years after Darwin, after the recognition of the variability in nature and of gradual transitions rather than fixed and sharp borders, after the recognition of variation and change as fundamental features of social forms, human institutions, language and cultures, that the thinking of the clinical science shows so much reluctance to give up the old ideals of exact definitions of sharply divided disease entities, and explanations in terms of necessary or sufficient conditions?

(5) In the present paper attention will be focussed on the problem of defining diseases. Without in any way reducing human disease to biological species, an analogy drawn from the methodology of evolutionary biology appears to be very illuminating in relation to our problem. In post-Darwinian biology a distinction has to be made between species as units of classification and species as units of evolution.[1] The species in the latter sense is a part of the fabric of nature: collectives characterized by a specific relation (inter-breeding with fertile offspring). A unit of classification is a human construct, a class of organisms delimited by the taxonomist by applying a prototype as a standard of classification. By applying that categorical distinction, the Feinsteinian [2] intuitions of diseases as evolving entities can be made precise.

Disease entities are units of classification. As such they can be defined (by listing the characteristics of the prototype) and as such they are not evolving and not characterized by variability. Disease entities are, however, *abstractions* from a variable, evolving *disease kind*.

The Feinsteinian Venn diagram representation of syndromes is seen as a way of representing the variability of a disease kind and partly representing the evolving characteristics of the disease kind by specifying the differing prognosis contained in different subsets of the diagram. In the light of this, syndromes (represented in the Feinsteinian manner) should not be regarded as a primitive and preliminary way of characterizing human disease, as is often the case; they cannot be replaced by well-defined disease entities.

This view has, if true, also important consequences for the problem of causality in medicine. The vocabulary of necessary and sufficient conditions is, as mentioned earlier, a part of the vocabulary by which we specify (well-defined) disease entities. If the focus of clinical practice and theory of the future is to be on variable syndromes rather than specific entities, then the discussion of causality in medicine must change its vocabulary.

2. CAN DISEASES BE DEFINED?

(6) In Wulff's *Rational Diagnosis and Treatment* ([11], p. 67), definitions are considered as crucial for a rational clinical practice. In other discussions definitions are treated with less deference. Strömgren ([9], p. 787) claims that the concept of psychosis covers an important "reality" for everyone working in clinical psychiatry – though no definition of the concept is at hand.

Another psychiatrist, Villars Lunn ([6], p. 1265) agrees with Strömgren. He lists a number of characteristics of the psychotic patient. But, he says apologetically, it is not a real definition. It is not clear from the list which characteristics are necessary (respectively, sufficient) for determining the psychotic phenomenon. This is, according to Lunn an *inconsistency* which, however, is necessary if the characterization is to have a certain breadth.

Lunn's position raises certain problems. In what sense is a concept inconsistent if we are not able to specify necessary and sufficient conditions for something to fall under the concept? And how can an inconsistency be justified at all?

(7) Wulff apparently maintains the strong conception of definitions: ". . . the defining criteria must fulfill the requirement that they are present in all patients suffering from the disease and absent in all patients not suffering from the disease" ([11], p. 67). What is sometimes presented in textbooks as a definition of disease is often on closer examination found "not to be a logically satisfactory definition but only an ultrashort description" ([11], p. 50).

But why is it important to define diseases? Wulff is not arguing in an epistemological way. He is not claiming that the clinical sciences should define their material because philosophers have insisted on that ideal. He argues instead from a pragmatic point of view: "A clinician cannot utilize the experience which another clinician has acquired by treating patients with, for example, duodenal ulcer or Crohn's disease of the colon, if he does not diagnose the diseases in similar patients" ([11], p. 51). We need definitions of diseases to facilitate communication between clinicians, and to make possible collective utilization of the experience of individual clinicians.

Definitions are thus seen as preconditions for a rational practice. But how do we obtain the definitions? How do we decide whether a definition is correct or not?

According to Wulff "all definitions are arbitrary". It is meaningless to discuss whether they are true or false. So, though Wulff sticks to the definitional ideal, he dissociates himself from the essentialist tradition in medicine.

(8) If definitions cannot be said to be true or false, correct or incorrect, how do we decide at all whether we should accept a proposed definition? Or to put the question in another way: Has Wulff simply not given up the strong conception of definition without making clear the logical character of the conception of definition to which he appeals?

Let us assume that the definition of diseases cannot be true or false. By that, Wulff is implying that to define a disease is not to specify characteristics which separately are necessary, and together sufficient, for the existence of a disease of the kind. (Contrary to the position quoted above). A closer look at Wulff's reasoning shows, however, how he tries to find a path between accepting a strong conception of definition and regarding the definition as being a result of arbitrary stipulations.

Wulff claims in another context that the criteria "are fulfilled by all patients *said* to be suffering from the disease" ([11], p. 50). In the Danish edition of *Rationel Klinik* [10], Wulff does not use this expression. Here he just claims that the criteria are fulfilled by all patients *suffering* from the disease ([10], p. 65).

In other words: in usual medical practice, certain patients receive the same diagnosis. Defining the disease which is ascribed to patients by using that diagnosis consists of enumerating certain characteristics common to those patients.

Do we not face a vicious circle here? A rational clinical practice is said to *presuppose* defined diagnoses, but on the other hand defining the disease presupposes a practice where diseases are diagnosed.

(9) The problem can be presented in the following manner: Wulff is really (to put it in a philosophical way) trying to make clear the conditions which must obtain for saying that someone (*S*) 'has the concept of disease *D*'. In one context he seems to imply that the ability of *S* to *define* *D* is the condition for his having a concept of *D*. The concept as definition: C_d.) Two clinicians who *define* a disease in different ways thus have different concepts of the disease and accordingly treat it in different ways. In another context

he seems to imply that the ability *to perform certain tasks* (say diagnosing and treating D) is the condition for having a concept of D. (The concept as practical abilities: C_p.)

(10) C_d and C_p may appear as two conflicting conceptions of what it is to have a concept. We shall, on the contrary, argue that C_d *presupposes* C_p (not only in the temporal sense that verbal definitions *come* after the aquisition of certain practical abilities, but in the stronger sense that the domain of validity of the definition can only be delimited on the basis of C_p).

A collective carries out a certain practice in accordance with certain standard procedures for handling the tasks specific to the practice. Being able to act in accordance with a standard procedure does not *presuppose* an ability to analyse the procedure (i.e., to give a specific definition of the situation where the procedure is adequate). I know what the typical procedures are in typical situations. The clinician knows what the typical procedures are when a typical case is presented. Having a C_d of a disease is being able to *describe* the practice (in which C_p is realised). That means C_p is a precondition for C_d.

(11) We can illustrate that relationship between C_p and C_d by analyzing Villars Lunn's discussion of the concept of psychosis. Lunn lists a number of characteristics, but he does not pretend to have given a definition of the concept of psychosis. He makes it clear that it cannot be seen from the list to what extent the specific elements are necessary (respectively, sufficient) for determining psychosis.

But is the list quite arbitrary? The answer is no. It is really a description *of standard situations* from psychiatric practice. The list is thus an exact *definition* — not of psychosis in general (whatever that means) — but of psychosis as presented in standard situations (in a specific practice). Saying that the list is not an exact definition of psychosis in general could (but should not) be understood as an apology for the backwardness of psychiatry. It is a consequence of the fact that psychosis is a kind of phenomenon with such a degree of variability that the characteristics of standard cases are not shared by all cases of the disease.

There is variation within a kind of disease. Under different geographical and historical conditions a disease manifests itself in different ways. There is no common essence that can be specified for a specific kind of disease. If definitions were specifications of trans-historic, trans-geographic properties, there would be no definitions of disease. There is, however, a relative stability (under limited historical and geographical conditions) of manifestations of a disease. Therefore we can establish standard procedures for standard situations. Definitions are relative to such procedures and situations. On the other hand, reflecting on and describing the procedures and standard situations (i.e., defining the disease relative to these conditions) makes it possible to delimit the domain of validity for specific procedures and so to criticize the application of procedures outside their domain.

(12) Clearly, it is very important to stress the concrete practical presuppositions of diagnostic criteria, that they are always bound to a specific situation where typical cases are handled with procedures determined by the available means of investigation and treatment. Giving the *reasons* for the diagnostic criteria used in a given situation is then giving the reasons for exactly *these* procedures being at hand and being useful in the situation. In that way, it is possible to evade the problems Wulff faces when he says that to define a disease is to fix a set of criteria which are fulfilled by all patients *said* to be suffering from the disease.

According to that formulation, pure institutional or social *conventions* (for excluding certain groups) would suffice to delimit a disease entity.

(13) Nothing in the above analysis contradicts Wulff's idea that it is important for the communication of clinical experience to know how other clinicians use a specific disease term, and that this can be ensured by defining the terms.

It is, however, important to stress that defining the term is not a precondition for the term having a meaning within a clinical practice. It has a meaning if it is used in accordance with standard procedures of the practice.

It is even more important to stress the following: the definition of a disease term will not satisfy the essentialist criteria of exact definitions (while it is always relative to a specific practice and its standard procedures). This does not, however, imply that the definition is arbitrary. It is true — relative to a concrete practice.

(14) Diagnoses are not introduced on the basis of strong definitions, nor are they pure conventions. But how are they introduced in a practice? Can we take away the touch of arbitrariness simply by letting the diagnoses rest on standard situations and standard procedures? Lunn's list of characteristics presupposes, as argued above, standard cases of psychosis. Exactly these standard cases show, however, what seems to be a typical trait of traditional clinical diagnoses: that the standard cases are persons who, in the situation where they serve as standard, are (or in a future situation will be) unable to cope with normal social life, they are invalids in relation to normal challenges of life (in extreme cases they are unable to live).

It is important to have these roots of clinical diagnosis in mind today when we so often discuss the question of over-diagnosing and excessive treatment. We are seeking safeguards against these unintended consequences of the modern health system. There is today in many cases no trivial or evident answer to the question "why shall x be treated?" If we go back to the birth of the clinic (to Sydenham, e.g.) the situation was quite different. Diseases had, *as threats to the human life*, to be treated.

(15) Clinical standard cases are the presuppositions of diagnostic categories. Cases of the standard type are referred to by the diagnostic term. Cases which in some respect resemble the standard case are referred to by the diagnostic term. In every clinical practice there are procedures for negotiation and for settling if a concrete case is sufficiently similar to the standard case to be labelled by the diagnostic term. There is, however, no decision procedure (no definition in the strong sense) and this is, of course, one of the explanations of why there has been, during the development of clinical practice, a continuous development of *test standards*. There has been a continuous production of new tools for delimiting the cases to be treated.

It would here be quite misleading to interpret the test standard as a means of *redefining* or *replacing* the original diagnosis. The introduction of a sereological diagnostic test for rheumatoid arthritis did not furnish the clinic with a definition of the disease. The test standard is secondary in relation to the original standards of the disease in the following sense: a negative result of the test would not make the clinician reject the diagnosis if the clinical evidence were strong (i.e., if it were a clinical standard case). The original clinical standard is also a presupposition of the test standard in another sense: it is precisely the function of the test standard to justify which *non-typical* cases are to be labelled by the diagnosis and which are not.

3. REPRESENTING THE VARIABILITY OF DISEASE KINDS BY SYNDROMES

(16) It has been argued that no definition can cover all the variable cases of a kind of disease. Two questions, one theoretical and one practical, must still be answered.

(a) What is the justification for talking about different *kinds* of diseases if the kind cannot be defined? Is it not at least necessary to assume that it will be possible at some future time to reveal the common (now hidden) structure of the kind? — that, in a scientific medicine of the future, it will be possible to describe for example biochemical and genetic traits common to all cases of disease?

(b) Is this assumption not a necessary condition for developing rational strategies in clinical practice? It may be uncertain today whether an individual case belongs to a specific kind. But is it not a necessary condition for trying to act rationally (basing my activities on prior experience) to assume that the case belongs to a specific kind with certain specific properties?

We shall only make some brief comments on the second question: If our practice is directed towards objects which change and are variable, it would be irrational to act in that world as if it were a world of objects with unchanging essences. It is our opinion that one of Feinstein's [2] most important theoretical contributions has been to stress and argue in terms of this position. To argue that many shortcomings of modern medicine are due to an unfortunate development of clinical science during the last hundred years, a clinical science which has developed clinical strategies towards entities which are not the real variable developing diseases of the clinic, but are rather still photographs or time slices produced in the laboratory. This brings us to the first of the questions raised above.

(17) The idea that a kind is characterized by a real *essence* (common to all members of the kind) is deeply rooted in some mainstreams of our philosophical tradition and in the tradition of pre-Darwinian biology. In taxonomy, sophisticated systems of classification were developed. One of the requirements of a system of classification was that division of a domain into classes should be exclusive and exhaustive. The strong conception of definition was, in this way, built into the system. In general, systems of classification were interpreted as representing the structure of reality. The classes of the system were conceived of as representing *natural kinds*.

The Darwinian breakthrough in biology and modern evolutionary biology makes a more sound interpretation of the taxonomic classes possible. The natural species are not *classes* of individuals with a common essence. Species are collectives, characterized by great *variability*. We must distinguish between biological species (the units of evolution) and taxonomic species (essence, units of classification).

Units of classification *presuppose* units of evolution. By certain standard procedures (for example, selecting a type specimen of a concrete population of a species) the taxonomist constructs a class (of organisms in certain respects similar to the type). The unit of classification is defined by listing the set of characteristics shared by the members of the class. Units of evolution cannot be defined (because of geographical and historical variability).

(18) Using this distinction we can express the Feinsteinian lesson in the following way: characterizations of diseases established under specific experimental (standardized) conditions delimit *units of classification*. It is, however, necessary to distinguish between these units and the natural course of the disease (the disease as unit of evolution). Feinstein mentions a number of diseases "discovered" by the new methods of investigation in the basic medical science. He argues that "many of the new names, however, represented 'diseases' identified by laboratory workers, not by clinicians, and the diseases had been, if correlated at all, associated with isolated clinical states, not with prognosis" ([2], p. 120).

In that light *definitions* become unimportant, or even an obstacle to representing the disease or a unit of evolution. Wulff has many examples which support the Feinsteinian position, but he does not abandon the ideal of defining the diseases. He has, as already mentioned, no illusion of finding true definitions, but fruitful ones. A definition must serve its practical purpose and one of the criteria for that is, according to Wulff, that patients selected by using the diagnosis (preferably) "must form a homogeneous group as regards prognosis" ([11], p. 67).

The whole Feinsteinian technique of representing diseases by Venn diagrams presupposes, however, that this ideal cannot be realized. The diagram represents the *variation* within a specific kind of disease (say rheumatic fever). The diagram represents the variability in prognosis for patients in different parts of the spectrum. So the research goal is not to seek a definition of the disease, but to "select the categories to be cited in a prognostic classification of the specific clinical spectrum . . ." ([2], p. 150).

To select the categories to be cited in a prognostic classification of a specific spectrum requires "close clinical observation and follow-up data on large numbers of patients" ([2], p. 151).

(19) It is now possible to draw some conclusions concerning the concept of disease: We have to distinguish between diseases as *kinds* (of changing and variable cases) (units of evolution) and diseases as disease entities, i.e., specification of states (time-slices) of a disease (diseases as units of classification).

Specification of disease entities *presupposes* clinical test standards. In what sense are diseases functions of our systems of classification and mode of description? Wulff refers to diseases as vehicles of clinical experience. But the apparent non-reality of disease is not — as implied by Wulff — due to the lack of true (exact) definitions. Disease entities are unreal in the same sense in which taxonomic species are unreal. Taxonomic entities are *classes* and thus a product of our practical intervention. But the entity (the biological species or the evolving disease) which is made the object of our intervention is real and existing. In biology we have a general theory explaining the place of species in nature and specifying the mechanisms governing the evolution of species. We do not have such a theory in the clinical sciences explaining the structure and mechanisms of diseases as evolving entities. There are reasons for believing that no simple theory could perform that task. Change and variability within a kind of disease are not simply a function of biological mechanisms, but also of psychological and sociological mechanisms.

Kinds of disease are labelled units of *evolution, by analogy* with biological species. But — contrary to the essentialist tradition of Sydenham — we are not reducing kinds of disease to species. Such a reduction was in a sense reasonable in the pre-Darwinian era. Species, as well as diseases, were conceived of as units of classification. Now, because we are able to distinguish between units of classification and units of evolution, things become more complicated at the scientific level whilst at the same time we become more able to "tidy up" at the conceptual level.

(20) Is it possible to draw any conclusions for the strategy of clinical practice and research?

We shall point to one consequence concerning the role of syndromes in clinical science. In the modern discussion there has been a tendency to regard syndromes as a primitive or at least preliminary kind of disease specification.

Murphy talks of a syndrome as a "precarious entity", "in which the elusive facts are propping each other up by their association" ([8], p. 115).

Wulff seems (though he is not quite explicit in his presentation) to regard syndromes as subordinate to anatomically and causally defined diseases. If our arguments are correct, this position is not tenable, and besides it seems to contradict Feinstein's general point of view, which Wulff supports.

(21) To represent a syndrome in a Feinsteinian way is an attempt to represent diseases as units of evolution. Anatomic definitions or causal definitions resulting in "still photographs" of the disease or in specifications of parts of the spectrum may *contribute* to more realistic representations of the disease kind. But they are not rivals to syndromes. The syndromes of today cannot be replaced by causal definitions, but by more precise representations of the variations within the spectrum of the disease.

Basic research is an indispensable part of modern clinical science, but in a sense it should be regarded as subordinate to clinical research in diseases as evolving units, a task to be performed by the whole collective of doctors, nurses, therapists, and others in daily contact with sick persons.

Aarhus University,
Denmark

NOTE

1 I have stressed the importance of the distinction between units of evolution and units of classification in a number of other articles; see [3], [4], [5].

BIBLIOGRAPHY

[1] Engelhardt, H. T., Jr.: 1984, 'Clinical Problems and the Concept of Disease', in this volume pp. 27–41.
[2] Feinstein, A. R.: 1967, *Clinical Judgment*, (5th reprinting, 1976), Williams and Wilkins, Baltimore.
[3] Jensen, U. J.: 1981, 'Preconditions for Evolutionary Thinking' in U. J. Jensen and R. Harré (eds.), *Philosophy of Evolution*, The Harvester Press, Brighton.
[4] Jensen, U. J.: 1982, 'Theories, Historical Units or Units of Historical Reconstruction', *Czechoslovak Studies in the History of Science*, Prague 1982, pp. 385–413.
[5] Jensen, U. J.: 1983, 'Tradition und Repräsentation', in D. Henrich (ed.), *Kant oder Hegel?*, Klett-Cotta, Stuttgart.
[6] Lunn, V.: 1979, 'Psykosebegrebet', *Ugeskrift for Læger*, København.
[7] Mackie, J. L.: 1974, *The Cement of the Universe*, Clarendon Press, Oxford.
[8] Murphy, E. A.: 1976, *The Logic of Medicine*, Johns Hopkins University Press, Baltimore, Maryland.
[9] Strömgren, E.: 1969, 'Use and Abuse of Concepts in Psychiatry', *American Journal of Psychiatry*, 126, 6.
[10] Wulff, H.: 1973, *Rationel Klinik*, Munksgaards Forlag, København.
[11] Wulff, H.: 1976, *Rational Diagnosis and Treatment*, Blackwell Scientific Publications, Oxford, U.K.

HENRIK R. WULFF

COMMENTS ON JENSEN'S
'A CRITIQUE OF ESSENTIALISM IN MEDICINE'

The paper 'A Critique of Essentialism in Medicine' is a good illustration of the need for closer cooperation between philosophers and clinicians. Juul Jensen states that the establishment of a disease entity is not arbitrary, but that it is born of clinical practice. This point of view is very similar to the one expressed by Engelhardt at this meeting, that a disease entity represents a clinical problem, as current practice presupposes the ability to identify problems. However, a philosophical analysis of the concept of a disease entity does not solve the practical problems which face the clinician. Juul Jensen draws the traditional analogy between the disease classification and biological species, but there is a great difference.

I do not really mind that zoologists may be unable to define exclusively and exhaustively what is understood by an elephant, because if we let a group of zoologists loose in a zoological garden and ask them to point out the elephants, then I am quite sure that the inter-observer agreement will be high. If, on the other hand, we let a group of doctors loose at a hospital and ask them to point out the patients with pneumonia, schizophrenia, gastric ulcer, myocardial infarction etc., then we know that the inter-observer agreement will be far from perfect.

This fact has very important implications. I may read in a medical journal that 70% of gastric ulcer patients are cured, if I give them a certain treatment, whereas only 40% are cured, if I give them another treatment. However, I cannot use this information if the authors do not tell me how they define gastric ulcers. About half of all gastroenterologists include the so-called prepyloric ulcers whereas the other half excludes this ulcer type. Since prepyloric ulcers comprise more than 50% of all stomach ulcers, the problem is important.

I may also read in a well-known textbook of medicine that, say, 84% of patients with systemic lupus erythematosus (SLE) suffer from fever and than 92% have arthritis or arthralgia, but the practical use of this information is also limited, when the author does not tell me which patients he is talking about when he uses this diagnostic label. The disease classification serves to pidgeonhole all clinical knowledge and experience, and at present it is much too imprecise.

L. Nordenfelt and B.I.B. Lindahl (eds.), Health, Disease, and Causal Explanations in Medicine, 75–76.

We need working definitions of disease entities, and I like Popper's formulation that definitions must be read from right to left and not from left to right [1]. It is necessary to establish a definition of a disease like SLE, but the statement that SLE is defined by the symptoms and signs a, b, c and d, does not answer the question: What is SLE? It answers this question: What shall we call those patients who present a, b, c and d? It is interesting to analyse the origin of the disease classification, but we must not forget that it serves a practical purpose and that it must be made explicit.

Herlev University Hospital,
Denmark

BIBLIOGRAPHY

[1] Magee, B.: 1973, *Popper*, Fontana, London, U.K.

HELGE MALMGREN

PSYCHIATRIC CLASSIFICATION: THE STATUS OF SO-CALLED "DIAGNOSTIC CRITERIA"

I

Psychiatric classification and concept formation have long been the subject of considerable controversy. The importance of *operational diagnostic criteria* is nowadays often stressed ([8], [12]). It is argued that the traditional kind of disease descriptions in terms of "typical manifestations" be abandoned in favour of precise statements of what observable traits are required for the diagnosis in question. The recent American classification [5] is much influenced by this kind of "operationalistic" philosophy.

There are a lot of questions connected with the idea of operational criteria, which I am *not* going to discuss here, e.g., the problem of observability. In the following, I take for granted that, given a certain competence, equipment, etc. of the observer, it is possible to distinguish between observable and non-observable properties. Instead, I am going to concentrate on the following question: should a certain set of operational diagnostic criteria for a concept B be regarded as a *definition* of the concept B, in the usual philosophical sense of a set of analytically (or, in a wide sense, "logically") necessary and sufficient conditions for the application of the concept B, or should such diagnostic criteria be interpreted in a different way? This main question has three different parts:

(1) If operational criteria do not express analytically necessary and sufficient conditions, how do they function?

(2) What reasons are there for using operational criteria in the one or the other way?

(3) How do different authors use their proposed operational criteria?

In the following I am going to use the term 'Definition' (with capital 'D') to mean definition in the usual philosophical sense explained above. In non-philosophical texts, the term 'definition' is very often used in other senses (cf. [1]). In most medical textbooks of not too recent date, the heading "Definition" usually signals a concentrated description of the most important manifestations of the disease in question. It is usually evident that this description cannot be reformulated as a Definition in our sense. Therefore, even if a certain writer on psychiatric concept formation refers

77

L. Nordenfelt and B.I.B. Lindahl (eds.), Health, Disease, and Causal Explanations in Medicine, 77–87.
© 1984 *by D. Reidel Publishing Company.*

to his operational criteria as "definitions", one cannot be sure that what is meant is Definitions. Although there is some evidence that one of the main proponents of operational criteria in psychiatry [8] is really advocating operational Definitions, the status of the diagnostic criteria is much more unclear in other recent works, e.g., *DSM*-III (cf. [10]). With this I will for the moment leave the exegetical question (3), since I think that questions (1) and (2) have greater philosophical interest.

We now pass to question (1). In some contexts, operational criteria can of course have another function than that of logical conditions. Sometimes they are simply empirically derived probabilistic indicators of a disease process, which is Defined in a way logically independent of the criteria in question. The rales that can be heard in pulmonary oedema constitute an important probabilistic indicator of this condition, but the presence of rales is not part of the Definition of the concept 'pulmonary oedema' since this concept is Defined in pathophysiological terms. However, the situation is not so simple in the case of most psychiatric diseases. It is not the case that, e.g., *DSM*-III offers a number of Definitions *beside* the lists of diagnostic criteria. *It seems, after all, that the lists of criteria do have to fulfil the function of determining the meaning of the psychiatric disease names.* Therefore, we can reformulate our question (1) as follows:

(1') Can a list of operational criteria determine the meaning of a certain term without constituting an operational Definition? If the answer is "yes", how is that possible?

In the following, I will argue that the answer to the first part of this question is "yes". Concerning the second part of question (1'), I will plead for the so-called "realistic" theory of meaning, according to which universal terms may refer "directly" to unobservable properties or kinds without the mediation of any Definitions, a fortiori without the mediation of operational Definitions. Operational criteria instead play the role of *pointing out* the property or kind in question.

II

I am now going to analyse two concrete examples from recent somatic medical history; the discussion of these two examples will, I hope, make clear the content of the realistic theory while simultaneously giving some support to it. It will also cast some light on the function of traditional so-called "definitions" in medical textbooks. We will later have to discuss the relevance of the examples for the psychiatric field.

As our first illustration we choose a disease, now probably extinct, the etiology of which is still not known in detail. The disease is *encephalitis lethargica* or "European sleep sickness", which occurred in several epidemics during the 1910s and 1920s. (The classical text is von Economo [6]; for a modern review see [9], p. 411 ff.) The clinical manifestations of encephalitis lethargica were very variable, although the triad (a) influenza-like initial symptoms, (b) somnolence and (c) eye muscle pareses characterised many cases. Psychiatric manifestations were common, as were chronic neurological sequelae, namely, Parkinsonistic symptoms. No specific etiological agent was found, although the epidemiological data strongly support the hypothesis that the disease was caused by a virus. The last mentioned fact makes it impossible to Define the term "encephalitis lethargica" in terms of a specific microbial etiology. Nor did von Economo try to Define the term in such a way. It is quite evident that his hypothesis that the disease was caused by a diplostreptococcus ([6], p. 76) is an empirical hypothesis and not a Definition. The variable clinical picture makes it a priori improbable that encephalitis lethargica — as this disease has usually been delimited — could be operationally Defined in terms of observable symptoms and signs. Against this can be objected, that an operational Definition can be disjunctive in form: "symptoms *A* and *B or B* and *C or C* and *D*", or the like. However, one can find independent evidence to the effect that encephalitis lethargica (as usually conceived) cannot be operationally Defined. One important piece of evidence is the fact that it has been discussed whether encephalitis lethargica can occur in clinical disguises other than those already known, e.g., as a psychiatric disease without somatic symptoms ([9], p. 414; cf. also [6], p. 47).

We also have to discuss the possibility of Defining the disease in terms of pathology. Note first that the description of the typical pathology of encephalitis lethargica is mainly founded on severe, acute cases. This means that although von Economo was of the opinion that he had pin-pointed a pathognomonic and obligatory pathological sign of encephalitis lethargica ([6], p. 60), it is quite probable that he would have revised his pathological criteria had it turned out that they were not adequate in a number of less severe cases. This means that he probably did not use the pathological sign as a Definition of his disease. By the way, the modern opinion seems to be that it is not always possible to differentiate encephalitis lethargica from other kinds of encephalitis on pathological grounds alone (cf. [2], p. 163 ff.).

From the above we can conclude, that the term 'encephalitis lethargica', as traditionally used, probably has no Definition in terms of microbiology, clinical signs or pathological findings. In spite of this, the diagnosis "encephalitis lethargica" has been extensively used. Instead of concluding that it has been used without having a meaning, I propose that *it has been used as a name which refers directly to a hypothetical disease entity, which is unknown as to its intrinsic nature and only known by its manifestations.*

To see how such a use of a term is possible, let us look at the situation from an epistemological point of view. von Economo and later authors have observed a number of uncommon cases, the manifestations of which have been very variable but which have still shown a lot of similarities with each other (type family-likeness). The spatio-temporal relations between the cases have been such that the hypothesis of a common etiology is made probable. Note how important this epidemiological evidence is. The intrinsic characteristics of the single case do not constitute any strong reason for postulating a previously unknown disease entity. In an isolated case, other explanations in terms of atypical manifestations of previously known diseases would be more plausible. However, such explanations rapidly lose their plausibility when we meet an epidemic of uncommon cases. von Economo discusses ([6], p. 28) the theoretical possibility that his thirteen cases were all caused by the same agent as that causing poliomyelitis. However, von Economo argues, then the same infection ought to have caused a number of typical cases along with the atypical ones.

The above argument leads to the following conclusion: *we can sometimes be quite sure that we have to do with a previously unknown disease entity, without being able to formulate reliable clinical criteria for the single case.* Furthermore, even in the epidemiological context the single diagnoses may be less certain than the general hypothesis that the great majority of cases is caused by the new disease entity. Faced in this way with a new disease, we will soon need a name for it in order to be able to formulate the hypothesis that another patient has the very same disease. For this purpose, we evidently cannot Define the disease in terms of its clinical manifestation, simply because we are not certain which they are. *The best we can do is to give a comprehensive description of the observed cases and to explain our use of the term* − e.g., 'encephalitis lethargica' − *by stating that it refers to the process which most probably causes most of these cases. This explanation of the meaning of our term is not an operational Definition.* The relation between the observable manifestations and the underlying disease process has the character of an empirical hypothesis.

It is most desirable that this hypothesis is made as precise as possible. It may well take the form of a list of "diagnostic criteria", which criteria then have the character of probabilistic indicators of the hypothetical disease process. It is important to be aware of the provisional character of such lists of diagnostic criteria. In the course of investigating a new disease entity, the accumulating evidence often leads us to revise our hypotheses about the relations between process and manifestations. However, all the time we can use the presumed empirical connections in explaining the meaning of the name of the disease to other people, since by learning which are the probable manifestations of a disease one learns where and how to look for it. Such an explanation is analogous to telling a child that a crow is the kind of bird which he can see if he takes a look through the window. This is not equivalent to a Definition, since it is only as a matter of empirical fact that a crow can presently be seen through the window. Instead, it is a verbal substitute for *pointing* to the bird in question.

It is now time to introduce our second historical example. I have here chosen a very recent discovery, namely, the so-called *"Legionnaires' disease"*. The reason is that for a rather long time, this disease had the same status as encephalitis lethargica still has: it was a disease of unitary but unknown etiology. During two weeks in July-August 1976, nearly 200 participants in a congress in Philadelphia fell ill with high fever, cough and pneumonia. 29 people died. The epidemic was over almost as fast as it had begun. Intensive research efforts to unravel the etiology of the epidemic began. However, four months later this work had made almost no progress. All tests for previously known microorganisms were negative (except in a few cases, cf. below). Neither could any unusual chemical substance be found which could have caused an intoxication with the symptoms in question (cf. [3]). Now, all this constitutes strong evidence that the epidemic was caused by the same, previously unknown agent. This conclusion is based on the negative findings on the usual tests, in combination with epidemiological evidence. The importance of epidemiology must again be stressed. The clinical signs and symptoms which the congress delegates exhibited where rather unspecific, and most cases would certainly have passed without causing much stir, had they occurred in isolation.

A couple of months later the mystery was solved. There was by then convincing evidence for the hypothesis that the epidemic was caused by a previously unidentified bacterium (later it was to be called *Legionella pneumophila*). It is of great interest to study the methodology and concepts involved in the decisive investigations of the disease ([7], [11]). A new bacterium was isolated from a few cases of the epidemic. The other cases were tested with respect to a serological reaction against this bacterium. A number of control groups was also used. In order to delimit the groups which were to be the primary subjects of the investigation, certain criteria were formulated ([7], p. 1190):

A case was considered *Legionnaires' disease* if it met clinical and epidemiologic criteria. The clinical criteria required that a person have onset between July 1 and August 18, 1976, of an illness characterized by cough and fever (temperature of 38.9°C or higher) or any fever and chest X-ray evidence of pneumonia. To meet the epidemiologic criteria, a patient either had to have attended the American Legion Convention held July 21–24, 1976, in Philadelphia or had to have entered Hotel A between July 1 and the onset of the illness. A person was considered to have had *Broad Street pneumonia* if he met the clinical but not the epidemiologic criteria for Legionnaires' disease but had been within one block of Hotel A between July 1 and the onset of illness.

The investigation showed that more than 90% of those patients who fulfilled the criteria for Legionnaires' disease and from which adequate sera were taken, showed sero-conversion or were significantly seropositive against the new bacterium. 9 of 14 who fulfilled the criteria for Broad Street pneumonia showed the same result, while sero-conversion or significant seropositivity was unusual in the other control groups. This is indeed convincing evidence for the hypothesis that the new bacterium was the cause of the epidemic.

Let us now discuss how the authors use their "criteria" of Legionnaires' disease and Broad Street pneumonia. The authors formulate one of their conclusions as follows:

that the large majority of cases that met the criteria for Legionnaires' disease and for Broad Street pneumonia *were in fact the same disease* ([7], p. 1195, italics added).

Now, the authors refer to the criteria for Legionnaires' disease as

the clinical and epidemiologic *definition* of Legionnaires' disease ([7], p. 1192, italics added).

It is evident (cf. the *first* of the three quotations above) that the criteria of Legionnaires' disease and Broad Street pneumonia, respectively, are logically incompatible. How, then is it possible that several of the individuals with Legionnaires' disease and Broad Street pneumonia had the same disease (cf. the *second* quotation above)? It seems to me, that the most plausible explanation is that, in the *third* quotation above, the authors did not mean Definition by the term 'definition'. Instead, what they call a "definition" is a set of probabilistic indicators of the hypothetical disease entity lying behind the epidemic. Then it is not difficult to understand how the criteria for Broad Street pneumonia can indicate the same disease.

There is, however, another possibility. The authors may have intended the operational criteria for Legionnaires' disease and for Broad Street pneumonia as Definitions, and they regard the "same disease" which is referred to in the second quotation above as a *third* disease lying behind both Legionnaires' disease and Broad Street pneumonia. We can find some evidence against this alternative in a contemporary popular article [4]. Here, the leader of the investigation in question (David Fraser) is said to entertain the hypothesis that

some of the individuals who had Broad Street pneumonia will be shown to have had Legionnaires' disease ([4], p. 470).

Of course, the question of the correct interpretation in this case is not very important. Any scientist is free to use terms like 'Legionnaires' disease' in a sense in which it is analytically equivalent to a certain set of operational criteria, i.e., as operationally Defined by these criteria. However, then he must not forget that such operational Definitions are provisional, and that their use does not exclude the possibility of simultanously using terms which refer to hypothetical disease processes. Independently of the actual intention of Fraser *et al.* [7] we can certainly say that, in the first 4 months after the epidemic in Philadelphia, it was possible to use the term 'Legionnaires' disease' as a name of the hypothetical process which caused this epidemic, and to use operational criteria as probabilistic indicators of that process.

III

Maybe the reader agrees with our analysis of the two historical examples, but doubts their relevance for the debate about psychiatric concept formation. Haven't we been taught not to try to define psychiatric concepts in terms of etiological hypotheses? Firstly, it is said, in psychiatry we seldom find the kind of distinct causal connections which we meet in the context of infectious diseases. In psychiatry, we usually have to do with multifactorial etiology, both in the sense that several causal factors contribute to the shaping of the clinical picture in the single case, and in the sense that the same clinical picture can have different causes in different cases. Secondly, many doubt that categorical concepts are useful in psychiatry, and instead argue for dimensional concepts. There are no natural boundaries between, e.g., different types of schizophrenia in the way there are natural boundaries between different viruses. Finally, there is little theoretical agreement in today's psychiatry. Without such agreement, how could we possibly use disease concepts founded on theories about the causes of the observable manifestations?

In my opinion, only the third of these objections carries any weight. Let me first say a few words about the first and the second objection. In spite of what Kendell says ([8], p. 64 ff.), the concept of a 'disease entity' is not made useless by the fact that many causal factors interact to produce a certain clinical picture. von Economo, e.g., was not unaware that the clinical manifestations of encephalitis lethargica could be modified by factors other than the process he postulated (cf. [6], p. 31). But this is exactly the reason why he needs a name for the hidden process, as distinct from the clinical picture! If there were an unambiguous connection between the clinical picture and the underlying disease process, one could equally well define the disease concept on the symptomatic level or let it stand for the underlying process. When the symptom picture is influenced by many factors it is, on the other hand, as necessary to have a name for each of these causal factors as to be able to name the clinical picture as such. The question about categories versus dimensions is, I think, independent of our main discussion in this essay. One can use both dimensional and categorical concepts at the hypothetical level. Consider, e.g., the concept of an 'allergic disposition'. This disposition can be quantified, and it fades imperceptibly into normal variants. However, it is still distinct from other etiological factors behind allergic manifestations as, e.g., the degree of exposition to certain allergens.

The objection that there is not much theoretical agreement in today's psychiatry is more important. I don't want to underestimate the danger involved in using premature, etiologically based classification systems in psychiatry. However, I want to bring three factors to attention in this discussion. First, one must distinguish between on the one hand, hypotheses about *specific etiology*, and on the other, hypotheses about *unitary but unidentified* etiology. The concept of 'encephalitis lethargica' involves reference to the second kind of hypothesis only. The same was the case for Legionnaires' disease before the discovery that *Legionella* is the specific cause of this disease. Although it may presently be impossible to base psychiatric classifications on specific etiology (cf. premature concepts like 'frontal lobe syndrome'), this need not be the case for classifications based on unitary but unidentified causes. When investigating supposed toxic or infectious causes of disease, spatio-temporal closeness between cases plays an important part (cf. above). In psychiatry, other kinds of relationship are more useful in the search for unitary causes, e.g., biological kinship and similar physical or psychical stressors. There is also the possibility of using the longitudinal case study as a basis for hypotheses about unitary processes lying behind different manifestations, cf. e.g. manic depressive illness (and see below).

Secondly, agreement can be reached on different levels. Even if one cannot agree as to the remote causes of certain symptoms, the immediate mechanism behind the symptoms may be uncontroversial. Thirdly, it is not only a question of communicating existing knowledge. We also use concepts to formulate *new* hypotheses which can then be tested against observable data. In this process one can legitimately introduce tentative concepts standing for presumed underlying disease entities, in order to be able to test different hypotheses about these entities and their interactions. Let me go into some detail concerning the last two points. Even if there is much theoretical controversy in today's psychiatry, we ought to be able to co-exist peacefully on a low theoretical level. As an example I will take that characteristic disturbance of recent memory which goes under the different names of "Korsakow's amnesia", "primary amnesia" and "the amnestic syndrome" (cf. [13]). This syndrome is characterized by a disability to retain new material for more than the short time that the patient can keep his attention fixed on the same thought. Disorientation of course follows. The patient further has a tendency towards confabulation, which can be more or less pronounced. Intellectual functions other than recent memory are, at least in pure cases, amazingly well preserved. Korsakow's amnesia can occur together with other serious disturbances of mental functions, e.g., delirium. In this case, the primary amnesia can be impossible to recognize. This depends partly on the fact that delirium can itself lead to memory difficulties and disorientation, partly on the fact that it may be practically impossible to test memory function. However, it is not uncommon that a longstanding amnestic syndrome is complicated by a brief episode of delirium. The symptoms of the patient before and after this episode give us strong evidence for the hypothesis that the amnestic syndrome was there the whole time, even during the delirium, although the amnestic syndrome could not have been recognized if one had met the patient only during the delirium. Our formulations presuppose that we conceive of the amnestic syndrome as a hypothetical process which does not always manifest itself "on the surface". We cannot give this process an operational Definition which covers all possible situations and interactions with other processes. In *DSM*-III, however (see [5], p. 112 ff.), the diagnostic criteria for *Amnestic syndrome* state that this diagnosis is *not* to be used when, e.g., delirium is at hand. This is, I think, an expression of an extremely operationalistic philosophy which does not even accept hypothetical processes on a very low theoretical level.

What I finally have to say about the third point formulated above also has a bearing on our original question (2), i.e., the question of reasons for using operational criteria as Definitions and using them as probabilistic criteria, respectively. I have made clear above that I don't see any theoretical objection to using operational criteria as Definitions. However, the usefulness of such Definitions has been over-emphasized in the recent discussion, and I think that this may hamper scientific progress in psychiatry. In the hypothesis-forming phase of empirical science, the scientist is looking for structuring principles. He tries to find a simple underlying pattern which can bring order to the varying manifestations on the surface. The protagonists of operational Definitions argue that operational Definitions are of special value in this situation, i.e., before one has any plausible theories about underlying factors. I want to reply that operational Definitions can be especially dangerous in this situation, since they can bias the scientist in favour of given patterns on the observable level and lead him to overlook the more important patterns on higher, explanatory levels. Therefore, although the recent trend towards operational diagnostic criteria in psychiatry will have important positive effects on the replicability and communicability of scientific results, the sub-trend towards operational Definitions may well be an obstacle to theoretical progress.

University of Stockholm,
Sweden

ACKNOWLEDGEMENT

This work was supported by a grant from the Swedish Council for Research in the Humanities and Social Sciences (no. F 214/79).

I am much indebted to Kristina Malmgren and Göran Lindqvist for helpful criticism, and to Solveig Karlsson for the typing of the manuscript.

BIBLIOGRAPHY

[1] Achinstein, P.: 1968, *Concepts of Science*, Johns Hopkins Press, Baltimore, Maryland, and London, U.K.
[2] Anderson, W. A. D. and Scotti, T. M.: 1968, *Synopsis of Pathology*, 7th ed., Mosly, Saint Louis, Missouri.

[3] Culliton, B. J.: 1976, 'Legion Fever: Postmortem on an Investigation that Failed', *Science* **194**, 1025–7.

[4] Culliton, B. J.: 1977, 'Legion Fever: "Failed" Investigation May Be Successful After All', *Science* **195**, 469–70.

[5] *Diagnostic and Statistical Manual of Mental Disorders*, 1980, 3rd ed. (*DSM*-III), American Psychiatric Association, Washington, D.C.

[6] von Economo, C.: 1918, *Die Encephalitis lethargica*, Franz Deuticke, Leipzig, and Vienna.

[7] Fraser, D. W., *et al.*: 1977, 'Legionnaires' Disease: Description of an Epidemic', *New England Journal of Medicine* **297**, 1189–96.

[8] Kendell, R. E.: 1975, *The Role of Diagnosis in Psychiatry*, Blackwell, Oxford, U.K.

[9] Lisham, W. A.: 1978, *Organic Psychiatry*, Blackwell, Oxford, U.K.

[10] Malmgren, H. and Lindqvist, G.: 1982, 'The Status of the So-Called "Diagnostic Criteria" in *DSM*-III', unpublished.

[11] McDade, J. E., *et al.*: 1977, 'Legionnaires' Disease: Isolation of a Bacterium', *New England Journal of Medicine* **297**, 1197–1203.

[12] Spitzer, R. L., *et al.*: 1977, '*DSM*-III: Guiding Principles', in Rakoff, V. M., *et al.* (eds.), *Psychiatric Diagnosis*, Brunner/Mazel, New York.

[13] Victor, M.: 1969, 'The Amnestic Syndrome and Its Anatomical Basis', *Canadian Medical Association Journal* **24**, 1115–25.

STAFFAN NORELL

COMMENTS ON MALMGREN'S 'PSYCHIATRIC CLASSIFICATION: THE STATUS OF SO-CALLED "DIAGNOSTIC CRITERIA"'

Dr. Malmgren's paper raises the questions "How should a certain disease entity be defined?" and "How can we identify cases of a specific disease as distinct from other diseases?"

The paper deals with the status of diagnostic criteria in psychiatry, although the two examples discussed represent somatic and probably infectious diseases. We can come back to the question of whether or not these examples are relevant and representative of the situation in psychiatry (or for non-infectious somatic diseases) later.

But first I have to admit that it is easy for me to agree with many of the suggestions put forward in this paper:

(1) Disease descriptions in terms of "*typical manifestations*" are problematic from a scientific point of view.

(2) The use of operational *diagnostic criteria* will have important positive effects on the replicability and communicability of scientific results.

(3) The use of operational diagnostic criteria as *definitions* of disease entities may well be an obstacle to theoretical progress.

I think Helge Malmgren has argued quite convincingly that it should be possible to formulate empirical hypotheses concerning the relationship between observable manifestations and an underlying disease process. Hence, clinical manifestations or operational diagnostic criteria should not be used as definition of a disease.

How, then, should we define a disease? The answer to this question was not clear to me from the paper. On page 80 it is indicated that "the underlying process" should be used, and my immediate interpretation of this was *patho-physiology*. However, on page 84 I was given the impression that a future goal may be to base definitions on the *etiological* process. I am not sure whether this means that diseases should be defined by (1) patho-physiology, (2) etiology or (3) both.

Since I have devoted some time to testing empirical hypotheses concerning the etiology of certain diseases — and I hope to be able to continue to do so — I would prefer not to define diseases by their etiology (for the same reason as that given previously for not using diagnostic criteria as a definition).

Epidemiology obviously played an important role in the identification of

89

L. Nordenfelt and B.I.B. Lindahl (eds.), Health, Disease, and Causal Explanations in Medicine, 89–90.

the two disease entities described in the paper: encephalitis lethargica and Legionnaires' disease. In both cases the etiology was unknown — at least for some time. In both cases the clinical manifestations were unspecific or variable. In both cases descriptive epidemiological data showed a clustering of cases in time and also in their geographical distribution. This, of course, is a pattern well known from infectious disease epidemics. But my question here is: To what extent could we expect that variations in "time" and "place" distribution of cases will be useful to identify psychiatric diseases? My own feeling is that other descriptive epidemiological characteristics — such as variations in morbidity between different population groups — could be more useful in the identification of psychiatric diseases. Perhaps there are other kinds of data which are likely to be helpful to an epidemiologist looking for new psychiatric diseases.

Huddinge University Hospital,
Sweden

SECTION III

ON CAUSAL THINKING IN MEDICINE

A. CAUSAL ANALYSIS

ANDERS AHLBOM

CRITERIA OF CAUSAL ASSOCIATION IN EPIDEMIOLOGY *

Epidemiology may be defined as the science of occurrence of disease. One ultimate goal in this science is to detect causes of disease for the purpose of prevention. A discussion of the concept of causes is beyond the scope of this presentation. It suffices to note that the objective of epidemiology implies that, to be of practical use, causes must be such that their elimination has a positive effect on the disease occurrence, i.e., that it decreases the incidence rates. Obviously this means that epidemiologic research is focused on so-called contributing or component, rather than necessary or sufficient, causes.

The formal objective of an epidemiological investigation is usually the estimation of the magnitude of the effect (if any) of some exposure to some disease. The effect is then measured on some appropriate scale such as the relative risk, i.e., the ratio of the risk for those exposed to that for those unexposed. It is important to emphasize that the objective is to assess the strength of a causal association and not the strength of an association that may be either causal or non-causal. If a study, for some reason, shows an association between exposure and disease that does not reflect a causal relationship, the study result is invalid. This lack of validity may be due either to an association in the population that is non-causal in nature and which the study design has failed to control, or to a defect in the study design itself that generates such associations. For an example of the first type, consider the association between the habit of carrying match-boxes and lung cancer. That is clearly a non-causal relationship existing already in the target population. For an example of the second type, consider a study aimed at assessing the effect of smoking upon chronic bronchitis. Since smoking, is more or less, part of the diagnostic criteria for chronic bronchitis, a study that does not employ carefully chosen diagnostic criteria tends to heavily over-estimate the effect.

In principle there are three types of epidemiological studies that may be used to study the effect of exposure upon disease. As will be discussed later these also represent three levels of sophistication. First of all there are studies based on aggregate data, such as international comparisons of frequencies of exposure and occurrence of disease. These types of study broadly coincide with what are usually called descriptive studies. The second type of study

93

L. Nordenfelt and B.I.B. Lindahl (eds.), Health, Disease, and Causal Explanations in Medicine, 93–98.
© 1984 by D. Reidel Publishing Company.

is based on individual data describing the status of exposure and disease, with no influence of the investigator on the distribution of exposure. These studies are either cohort or case-control studies. In a cohort study, one group is exposed to a risk factor and compared, regarding disease occurrence, with another group not thus exposed. By contrast, a case-control study consists of a comparison with regard to exposure of a group with the disease (cases) and a group without the disease (controls). These types of studies are often called analytical or etiological studies. On the third level we find the so-called interventive or experimental studies. These are characterized by a random distribution of individuals on one group that is exposed and one that remains unexposed.

Regardless of what sort of study has been conducted, a demonstrated excess risk among those exposed may be due to one or more of the following [3]:
— causation
— randomness
— lack of validity
    — in the comparison
    — in the selection of index or reference group
    — in the measurements

Causation is what we are searching for. From initial planning to interpretation of collected information we are trying to rule out the influence of the other factors. Of those other factors, randomness is least difficult to cope with. The reason is that there is a widely accepted technology that is developed to tackle this problem. This technology is the general knowledge about statistical inference. Basically this methodology offers a means to assess the magnitude of random influence on the data. Even though there may be some issues that are under debate in this field there seems on the whole to be general agreement among biostatisticians and epidemiologists on how this technology should be used and how the results should be interpreted. Therefore I will not discuss the problem of randomness any more in this presentation.

The main threat to the validity of the study result consists of the three last factors on the list, i.e., lack of validity in comparisons, selections, and measurements. Consequently the main concern during the planning and conducting of a study must be to prevent any of these from influencing the result. Similarly the main issue to be addressed in the data interpretation must be to what extent the efforts to prevent any influence from these factors have been successful. The rest of this presentation is aimed at a discussion of how this should be done.

First I will try to illustrate the nature of the three sources of validity problems by means of some examples. For an example of invalid comparisons consider again the example discussed earlier dealing with the association between match-box carrying and lung cancer. In that example a comparison was made between people who carry match boxes and people who do not regarding the occurrence of lung cancer. This comparison is said to be of low validity since match box carriers comprise a greater proportion of smokers than the other group. The keypoint here is that smoking is a cause of lung cancer, and acts as a so-called confounding factor in this example.

Consider then the example of smoking and chronic bronchitis. Suppose that we are going to compare people with and without chronic bronchitis with respect to their smoking habits. However, since smoking may be used as part of the diagnostic criteria for this disease, the exposure that we are going to observe influences the selection of people in the group with chronic bronchitis. Therefore this selection is said to be of low validity.

For an example of invalid measurements let us say instead that we are comparing smokers and non-smokers with regard to the occurrence of chronic bronchitis. In this setting the problem with the diagnostic criteria would result in invalid measurements, since when we measure the occurrence of chronic bronchitis, people in the smoking group would more easily be assigned this diagnosis.

It seems that the criteria to be used for judging whether an association is causal or not may be broadly divided into three main groups [1] :
— Use of accepted scientific principles
— Biologic plausibility
— Statistical properties of the data

Let us first consider the use of scientific principles. This has to do with the design and performance of the study and can be examined without looking at the empirical data that have been collected in the study. The first thing to be considered is the basic design of the study. The first type of study discussed above was that based on aggregate data. Usually we would regard results from that sort of study as less valid than data from the other two kinds of study. The reason for this is obviously that this sort of study is difficult to protect from being influenced by the different sources of invalidity discussed above. Let us say for example that an international comparison displays an association between the sale of oral contraceptives and the occurrence of myocardial infarctions. Obviously it is extremely difficult to control for all possible confounding factors, such as food habits, which tend to make the comparisons of low validity, and also to control for other validity problems. Consequently, one should not decide from this type of data only that the use of oral contraceptives causes myocardial infarction.

In contrast, the randomized type of study is designed in a way that effectively decreases the problems connected with validity in comparisons, and also the problems connected with the validity problems in selecting the groups to be compared. Results from randomized studies are usually considered to be highly accurate at least as far as the randomization does not conflict with some other quality aspects of the study.

The problem for epidemiologists is that randomization can very rarely be performed in epidemiologic studies. Otherwise randomization would probably have been required as a scientific standard, as it has in most clinical trials, for instance.

The typical epidemiological study, however, is of the second type described above, i.e., a study with disease and exposure data on an individual level, but without randomization. Therefore the rest of this discussion will deal with cohort and case-control studies, which are the two study designs within this group. First of all there is the question of whether one of these two types of studies gives more reliable results than the other. There seems to be no general agreement among epidemiologists that this would be the case. Nor can the arguments that have been raised in favour or against one or other design be interpreted so that generally one study design is more liable to generate invalid results than the other. Therefore, whether the study is a cohort or a case-control study one has to consider each of the possible sources of validity problems equally carefully.

When reviewing the results of a study one has to check by which means and to what extent the sources of lack of validity have been tackled. Reading a study report, of course, the first prerequisite is that these problems have been acknowledged and that the ways they have been tackled are presented. For details about what should be required in these respects, reference is made to general literature about epidemiological principles [3].

The two other groups of criteria to be considered before deciding whether or not an association is causal are biological plausibility and statistical properties, respectively. Several authors have formulated criteria that fall within these two groups. Perhaps Bradford Hill's list of nine criteria is the most well known [2]. His list contains the following items:

(1) Strength
(2) Consistency
(3) Specificity
(4) Temporality

(5) Biological gradient
(6) Plausibility
(7) Coherence
(8) Experiment
(9) Analogy

By temporality Hill means that the exposure occurs prior to the disease. That criterion apparently is necessary for an association to be considered causal. This is the only absolute criterion on Hill's list. All other items on the list are relative in nature and have to be weighed against each other. I will not comment upon all items on Bradford Hill's list here, but just on a few of them.

Let us first consider some of the arguments belonging to what we previously called biological plausibility. If for example a certain substance has been shown to be harmful in animal experiments this is obviously strong support for causation if an association has been displayed in a study in man. Similarly, if a number of previous studies on different materials and with different technologies show the same result as you have found in a study, this favours a causal interpretation.

Finally, we shall discuss some criteria belonging to the group of criteria that we called statistical properties [4]. The most important elements within this group seem to be strength and dose-response relationship. The reason for using strength of an association as a criterion for causality is that if the association was due to confounding rather than to causation, then the effect of the confounding factor on the disease must be even higher than that which has been detected for exposure and disease. The idea is that if the association is strong enough a potential confounder would have to be so closely related to disease that it is unlikely that it has been overlooked. On the other hand, of course, this is not an absolute criterion since you can easily find examples of strong associations that apparently are due to confounding. Take again the association between match box carrying and lung cancer. This association is probably very high but nevertheless is due to confounding from smoking rather than being due to causation. It is equally apparent that a weak association may very well be due to causation and may of course not be rejected simply because of its weakness.

The properties of the dose-response criterion are similar to those of the strength criterion. When present, it seems reasonable to interpret a dose-response relationship as telling in favour of causation. On the other hand, confounded associations may show dose-response patterns as well. For example, match box carrying probably shows a dose-response relationship to lung cancer, since the more often you carry match boxes the more you probably smoke. Furthermore, associations may very well be causal without being related in the dose-response way.

In conclusion, the final decision as to whether an association should be regarded as causal or not requires weighing of criteria from the three groups discussed above. It seems that this weighing will necessarily contain a great deal of subjective opinion. Personally I would put most weight on the quality aspects of the study.

Huddinge University Hospital,
Sweden

NOTE

* This work was supported by the Swedish Medical Research Council (Project No. 6216).

BIBLIOGRAPHY

[1] Feinstein, A. R.: 1979, 'Clinical Biostatistics XLVII, Scientific Standards vs. Statistical Associations and Biological Logic in the Analysis of Causation', *Clinical Pharmacology and Therapeutics* 25, 481–492.
[2] Hill, A. B.: 1965, 'The Environment and Disease, Association or Causation', *Proceedings of the Royal Society of Medicine* 58, 295–300.
[3] Miettinen, O.: 1978, *Principles of Epidemiologic Research, Problem Conceptualization and Study Design*, Harvard School of Public Health, Boston, Massachusetts, manuscript.
[4] Rothman, K. J.: 1982, 'Chapter 2, Causation and Causal Inference' in B. Schottenfeld and J. Fraumeni (eds.), *Cancer Epidemiology and Prevention*, W. B. Sanders Company, La Follette, Tennessee.

GERMUND HESSLOW

COMMENTS ON AHLBOM'S 'CRITERIA OF CAUSAL ASSOCIATION IN EPIDEMIOLOGY'

I have very little to say about Dr. Ahlbom's paper by way of criticism, since I agree with almost everything he is saying. My only complaint is the very weak one, that Dr. Ahlbom has not succeeded in providing what no-one else has provided either, namely a theoretical underpinning of the arguments for and against causal hypotheses.

It is something of a scandal in philosophy of science, much more scandalous than the sheer *inability* to solve certain problems, that philosophers dealing with confirmation and induction have so persistently avoided the kinds of reasoning actually employed by scientists. A couple of years ago, when preparing a course in methodology for medical research students, I wanted to know what philosophers had had to say about the notion of scientific 'control' in the context of control group experiments. I searched through the indexes of some twenty standard books by well-known philosophers of science like Braithwaite, Popper, Carnap, Hempel, Nagel, etc., but not one of them contained any reference to the concept of control. Now, as everyone with the slightest acquaintance with empirical sciences like medicine, biology or psychology knows, control is *the* most important methodological concept. It is no wonder that scientists have not felt very enlightened by what philosophers have had to say here.

To illustrate what I mean about the lack of theoretical underpinnings of arguments about causal relations, I would like to draw your attention to two specific issues. Firstly, we may consider the list Dr. Ahlbom gives us on p. 94, where he considers the possible explanations for an increased risk. Although it ostensibly concerns the probability of contracting diseases, similar lists are common in methodological textbooks in the behavioral sciences, and I think that Dr. Ahlbom would agree that his list would be equally valid in any other empirical field and for any observed statistical association.

Now what I would like to know is whether this list is exhaustive of possible explanations. Suppose that I have good reasons for eliminating every alternative but the causal one; is it then logically inconceivable that the association is still not causal? My personal feeling is that it is, but I have no better argument for this belief than subjective intuition. Clearly, it would

L. Nordenfelt and B. I. B. Lindahl (eds.), Health, Disease, and Causal Explanations in Medicine, 99–100.
© 1984 *by D. Reidel Publishing Company.*

be preferable to have a formal argument showing that every logically valid objection to a causal interpretation must fall into one of Dr. Ahlbom's categories.

The second challenging problem mentioned by Dr. Ahlbom concerns the relative weight we put on arguments of different kinds. Say, for instance, that we have very good empirical evidence for the hypothesis that *A* causes *B*, but that there are also very strong theoretical arguments against it. What weight should we give to these arguments, and could a choice between them be defended? Dr. Ahlbom seems somewhat pessimistic here, for at the end of his paper he only mentions 'subjective factors' as possible guides. The only approach in current philosophical thinking, which would have anything to say on this subject is the Bayesian one, but Bayesianism achieves its comprehensiveness partly at the cost of concreteness. In practice I think that the Bayesian view of how to weigh the empirical and theoretical arguments would reduce to Ahlbom's 'subjective factors'.

Both of these problems are absolutely central to scientific practice but they have been virtually ignored by philosophers. It is high time this situation were changed.

University of Lund,
Sweden

ANNE M. FAGOT

ABOUT CAUSATION IN MEDICINE:
SOME SHORTCOMINGS OF A PROBABILISTIC
ACCOUNT OF CAUSAL EXPLANATIONS

This paper is a reflection on the shortcomings of a probabilistic account of causal explanations, and an attempt to comment on such limitations. A brief introduction serves to outline the main features of a probabilistic account of causality.

A Cause: A cause is a factor which raises the probability of an effect E significantly above its a priori probability (a factor which 'makes a difference') — a C such that $P(E/C) > P(E)$. Epidemiologists speak of *risk factors* and measure the degree of the causal *influence* through a correlation analysis. The clinicians' vocabulary includes a variety of nuances, basically: *external causes* (bacterial or viral agents, environmental pathogenic factors) vs. *internal causes* (predisposition, or diathesis: blood group, tissue group, etc.). The clinicians' causal judgments imply probability estimates for singular cases.

The probabilistic (Bayesian) theory may be understood as a theory of *efficient* causality. It implies that an (efficient) cause can never be a *necessary and sufficient* condition (that the necessary and sufficient condition case is a limit case which never materializes in practice). (See [7], [13].)

Etiological Diagnosis: An etiological diagnosis includes:

(a) a set of causal hypotheses $H_1, H_2, \ldots, H_n$ (exhaustive, and as far as possible mutually exclusive), an a priori probability distribution over the set of hypotheses (*initial conjecture*): $P(H_i)$ = a priori probability of H_i;

(b) a sequence of observations $E_1, E_2, \ldots$ (*data*) of the signs or symptoms offered by the patient, and for each E_k, $P(E_k/H_i)$ = the probability that sign E_k be observed in someone having disease H_i (sensitivity of sign E_k); and $P(\neg E_k/\neg H_i)$ = the probability that sign E_k be absent in someone not having disease H_i (specificity of sign E_k); i.e., the likelihood of H_i (medical knowledge, clinical or epidemiological); and

(c) a diagnostic inference (Bayes' theorem): at stage E_k,

$$P(H_i/E_k) = \frac{P(E_k/H_i) \cdot P(H_i)}{\Sigma_{j=1}^{j=n} P(E_k/H_j) \cdot P(H_j)},$$

where $P(H_i/E_k)$ is the a posteriori probability of hypothesis H_i, given observation E_k (*revised conjecture*). (See [4], [5], [6].)

101

L. Nordenfelt and B.I.B. Lindahl (eds.), Health, Disease, and Causal Explanations in Medicine, 101–126.
© *1984 by D. Reidel Publishing Company.*

The Cause. Specification of *the* cause involves a decision and forces the Bayesian procedure into a calculus of mathematical expectation. The Bayesian decision rule (*Bernoullian criterion*) stipulates that one should pick the hypothesis which has the highest expectation.

If it is assumed that committing an error carries a negligible risk (that selecting a false hypothesis, or rejecting a true hypothesis, has no unfavorable consequences; that there is plenty of time, etc.), then the hypothesis with the highest expectation is simply the hypothesis with the highest a posteriori probability: 'classical' decision rule. Pure research may conform to a classical decision rule.

If the consequences of error are taken to be serious, one should evaluate the desirabilities (as well as the probabilities) of the various hypotheses, and compute their expectations. The computation is often implicit. Thus, a sore throat in a child is treated as a streptococcal infection — even though another etiology may be thought *more probable* — because the streptococcal pharyngitis is *more serious*. In clinical medicine it is generally assumed that rejecting a true hypothesis (e.g., missing a diagnosis of acute appendicitis) is worse than accepting a false hypothesis (e.g., performing an unnecessary appendicectomy). In epidemiology one tries to maximize the informative content of hypotheses, with a view to stressing the importance of preventable or curable causes.

In the following I emphasize some (theoretical or practical) difficulties and limits of the probabilistic (Bayesian) model; most of them are also (nay even more) difficulties for the deductive-nomological account of causal explanation. Four aspects are analysed, about four case histories:

(1) Assessment of causal responsibility (case history *A*);
(2) Empirical etiological diagnosis (case history *B*);
(3) Pattern recognition vs. theoretical biases (case history *C*);
(4) Investigation of a new morbid entity (case history *D*).

1. ASSESSMENT OF CAUSAL RESPONSIBILITY: CASE HISTORY *A*

In May 1976, a young (unmarried) French woman, Ms. A, wanted an abortion (abortion had been legal in France since 1975). She consulted Dr. X. He agreed to take care of her (doctors have a right to refuse). They complied with all the formalities, and the operation was performed well within the legal delay (three months following the last period). The method used was the standard Karman aspiration technique. Blood clots and fragments were brought out, which Dr. X interpreted as being fragments of the placenta.

Two months later (July, 1976) Ms. A reckoned that she was still pregnant. The legal delay had elapsed. She went to London, where the pregnancy was successfully interrupted. Back in France, Ms. A brought an action against Dr. X for the prejudice she thought she had suffered, and claimed damages (F. F. 100 000 ≃ U.S. $20 000).

The court appointed a group of experts and put three questions to them:

Q1. What were the risks involved in the operation?

Q2. What were the possible causes of its being unsuccessful, i.e., did Dr. X meet the required technical standards?

Q3. Would there have been a means of checking whether the operation had been successful or not?

The experts answered:

A1. The main risk of the Karman method is a perforation of the uterus, a secondary risk is the possibility of an incomplete aspiration, resulting in hemorrhage or infection.

A2. The technique used in Ms. A's case is generally sufficient to interrupt the early course of pregnancy. However, in rare occasions, the aspiration may bring only blood clots and fragments of the uterus wall, which can be mistaken for fragments of the egg.

A3. In order to control whether the uterus is empty, it is possible to do a curetting (i.e., use a more effective and also more aggressive abortive technique). However it is neither compulsory nor common to do so.

The experts' report was found insufficiently enlightening, and the judges required additional information. The question was:

Q4. Did Dr. X fail to meet his duty to give the best treatment dictated by current scientific data and professional standards?

The experts answered:

A4. Dr. X failed to identify the exact nature of the fragments and blood clots he brought out. However, according to professional and scientific standards: he used an appropriate technique; he was not obliged to control the vacuity of the uterus through a more aggressive technique; in order to interpret the nature of the fragments correctly, he might have examined them under running water; but that is not a routine procedure.

Following the second report by the experts, the court settled on the following sentence: *Given that* physicians are not obliged to get the best result, but they are under the obligation to do their best and use all the means available to get the wanted result; *that* the experts reported that among the techniques available, Dr. X chose a milder and less performative technique; *that* the said technique may fail to draw the fetus away; *that*, in order to avoid any error of interpretation on the nature of the fragments obtained, it is scientifically established that the fragments can be examined under running water; *it is concluded that*: by using a technique the shortcoming and risk of which are known, and neglecting to take all the scientifically available steps for securing that the risk be infallibly excluded, Dr. X failed to meet his obligation to the means; he thus failed to satisfy to his contract with the patient. Hence he is to be considered responsible for the prejudice suffered.

Dr. X was cast in damages (F. F. 8000 ≃ U.S. $1600). The decision of the court was harshly criticized by some members of the legal profession. They claimed that the 'primary cause' of Ms. A's misfortune, necessary for her applying for abortion, was that she had intercourse with a partner of the opposite sex. Therefore it was not the ineffective medical procedure but the sexual intercourse which was the 'direct cause' of the prejudice suffered. (See *Le Concours Médical* **103** (1981), 6719–6720.)

Case history *A* illustrates a subjective difficulty people have thinking about causality in probabilistic terms. (It also shows how unreasonable and/or irrational it may be to stay away from the probabilistic approach.)

Fact: unsuccessful abortion, identified with a prejudice suffered by Ms. A. Wanted: the cause of the prejudice, i.e., of the unsuccessful abortion. Parenthetically, the French jurisprudence has so far tended to deny that the birth of an unwanted child, following medical failure to achieve effective abortion or tubal ligature, may be considered a prejudice in itself [7]. It may therefore seem paradoxical that in Ms. A's case a prolonged pregnancy terminating in a late and costly — but successful — abortion could count as a prejudice. That point will not be discussed here.

In their attempt at designating the cause, both parties assume the causal relation to be implicative. The Court wants the cause to be a condition *sufficient* to bring about the effect (i.e., they want the effect to be necessary, given the cause). The critics want the cause to be a *necessary* condition (i.e., they want the cause to be deductively inferable from the effect). The two points of view will be contemplated in succession.

Instead of pronouncing that implicit in the doctor-patient contract was the acceptance of the risk by both parties, the Court requires that the doctor should have taken scientifically dependable means in order to "infallibly exclude the risk". The thesis is carefully argued. In the normal course of things, an adequate technique provides effective abortion. The fact that abortion was not effective implies that there was something (the doctor's neglect) by which the normal cause was short of sufficient to bring about the effect. The flaw in the argument lies in the use of conflicting principles: (1) it is not the doctor's duty to get infallibly the wanted result; it is the doctor's duty to do his best and use whatever means are deemed technically appropriate by professional and scientific standards; (2) it is the doctor's duty to exclude infallibly the risk. But if the risk involved in the operation could be infallibly excluded by a competent technique, the doctor should be under the obligation to succeed infallibly.

What is perfectly well accepted from nature, namely, that its results are not automatic (e.g., to get pregnant; nature is not presumed mechanical, although it is presumed regular), seems more difficult to accept when a technical human action is involved. The fact that a diagnostic test is 'only' 90% specific, or yields 15% falsely negative results, is received reluctantly. Irregularities in therapeutic outcomes are held superable through methodological improvements. As the human performer is expected to be potentially omniscient and omnipotent, and failure is equated with incompetence, there is a corresponding professional tendency to overdo it, in order to secure the desired consequences, as in the recently reported case of a 40 year old man with carcinoma of the larynx. He underwent standard laryngectomy, with extensive bilateral removal of lymph nodes; then postoperative radiation, 6500 rads on the tumor area and superior cervical lymph nodes, followed by 5000 rads on the area of the inferior lymph nodes. The cancer did not recur, but after one year he developed post-radiation myelitis, ending in tetraplegia and death [14]. Surgery and radiation therapy had been sufficient indeed to eradicate the tumor — more than sufficient, one might say. Yet the optimistic belief that the harmful radiation is the radiation in excess, and that there always is a dose precisely sufficient to destroy the tumor without harming the patient, is unwarranted.

The causal factors we manipulate are rarely, if ever, 100% effective, and identically in all cases, with 0% side or detrimental effects. Ms. A's attending physician could have minimized the chances of survival for the fetus through using a more aggressive technique, or sending the fragments to a pathology department for more careful examination, or checking the vacuity of the uterus echographically. (There are examples of consultant experts or judges urging that either of those should have been done in similar cases [12]). However, strictly speaking, none of these precautions is strictly sufficient to secure the result, while they are superfluous in common cases. And one would hardly want Ms. A's doctor to have performed a hysterectomy, although that would more infallibly have excluded the risk that the fetus might survive.

It is certainly a problem to integrate the notions of accepted risk and fallibility without moving towards fatalism or laxity. However, (1) when the result is automatic, the causal question is rarely asked; (2) one should realize that there is also some degree of contingency in technical success (e.g., placebo effect of drugs). In one other abortion case (Rouen, 1982, in [7]), the Court sanctioned the doctor's neglect to warn the parents of the possibility that the operation might be unsuccessful. That seems fair. Note that in Ms. A's case the Court decided that the doctor's fault weighted only F. F. 8000 (instead of the claimed 100 000). The sentence was nevertheless found outrageous by some. Why?

Rather than arguing that it is impossible scientifically, nay infallibly, to exclude the risk (for the Court's decision is indeed questionable from that point of view), the critics had to fall on the woman's morals, and imply that *she* was responsible, because she should not have had sexual intercourse in the first place. The comment not only expresses a sexual prejudice, it also expresses a theoretical bias rather common among jurists. The cause should be a condition *sine qua non* [10]. Of course the doctor's mistake or neglect was not necessary for the abortion to fail. Abortion might have failed due to other causes, e.g., there were two fetuses, only one was aspired; or there was a major earthquake resulting in an electric shutdown making surgical interventions impossible for several months. Of course the sexual relationship was necessary. It is true that Ms. A would not have been in that distress if she had not had sexual intercourse. But it is also true that she would not have been in that distress if she had not been alive.

The critics presuppose that the causal relation is transitive (transitivity of implication): abortion was unsuccessful implies that there was an abortion, which implies that there was an unwanted pregnancy, which implies that she was pregnant, which implies that she had sexual intercourse. Finally, abortion was unsuccessful implies that she had sexual intercourse. Hence "sexual intercourse is the direct cause . . . ." But, if the critics were right, her being a female could be the "direct cause" of her distress as well, for, had she not been a female, she could not possibly have had an abortion.

Viewed as a necessary condition, a causal factor is such that, were it withdrawn, the effect could not materialize. When assessing causal responsibility, it is always useful to identify whoever could manipulate necessary conditions, i.e., prevent the occurrence of a detrimental effect. However, especially if the probability of the detrimental effect, given the 'necessary' cause, is low, claiming that the choice of allowing the causal factor to occur cannot be a rational choice is preposterous. You can be aware of the chance of a plane crash, and choose to fly to New York. If the plane does crash, people will say that, had you not taken that flight, you would not have died. They won't mean that you are causally responsible for your own death. Nor will insurance companies.

There is a medical and social tendency nowadays to be intolerant of heavy drinkers, cigarette smokers, motorcycle drivers, sportsmen, fatties, etc. — a tendency to make such people feel guilty for circumstances which may not quite be necessary conditions of illness, but of which it is thought that, had they been prevented, the person would not be ill. "Had you not eaten fat..." The present author remembers reading in a medical file the following judgement on a patient: "alcoholic intoxication, 50g/24h" (this amounting to half a bottle of wine per day). It may possibly be true that, had the man been strictly sober, he would not have contracted cirrhosis of the liver. It is certainly true that, had he not been born, he could not have had cirrhosis. The second condition is more necessary and less causal than the first. Necessary conditions are often trivial and uninformative causes. The causal link between cirrhosis and the consumption of alcohol is much more interesting. However, as it is probabilistic, the person may be held causally responsible for the choice of his drinking habits (to the extent that it was a choice), but he can hardly be blamed for the end result.

In fact, there may be a relation between the rigidity of ethical judgements (any failure or illness can be traced back to a misdeed) and the rigidity of theoretical analyses of causal links (a cause should be a condition sufficient, or necessary, to bring about the effect). Of course one might be tempted to settle on the conclusion that the cause should be *both* necessary and sufficient for its effect, but then the cause of Ms. A's misfortune has got to be Ms. A's misfortune itself — unless one believes in absolute determinism, in which case Ms. A's misfortune is a necessary consequence of any other state of the universe, and the precise question of causal responsibility does not have a meaning. The probabilistic view of causality is compatible with a more pragmatic and tolerant approach, but there is a problem of having an ideology of the liberty not to be perfect (the liberty to err) and recognizing the necessity of social and professional rules and obligations. When causal factors do not automatically produce or impede certain effects, it may become difficult to draw a line between ill-will and bad luck.

2. EMPIRICAL ETIOLOGICAL DIAGNOSIS: CASE HISTORY *B*

A 70 year old male, Mr. B, is hospitalized following an episode of hectic fever which set in abruptly two weeks earlier with a chill. His temperature has oscillated between 38° and 40° C ever since. The blood tests prescribed by the family doctor have shown an.elevated white cell count (leukocytosis: 11 000, 78% N.P.) and bacteriemia (blood culture: staphylococcus aureus with negative staphylocoagulase and positive Chapman reaction).

On admission, apart from the fever (40°C) and profuse perspiration, Mr. B appears to be remarkably well. He says he has always been in good health, and has no history of any chronic condition (in particular, no known allergy). He is thoroughly investigated and the findings are entirely negative.

The only clinical finding is a slight hypertrophy of the prostate, however causing no discomfort, and most probably benign.

A series of X-rays is performed; there may be an infected tooth (periapical granuloma of a canine); there is a suggestion of a mild irritable colon syndrome; the IUV confirms the overgrowth of the prostate and the absence of other lesions of the urinary tract; the rest is normal (chest, biliary tract . . .).

Laboratory findings: the blood syndrome is stationary (leukocytosis, elevated erythrocyte sedimentation rate).

The fever recedes spontaneously while the patient is at the hospital. The medical staff discusses the case and settles on the hypothesis of a staphylococcal infection, the primary focus of which has not been ascertained.

The patient is now apyretic. He is released from the hospital with a letter for his family doctor recommending that an eye be kept on him.

As soon as Mr. B has returned home, his temperature rises to 40° and the same pattern of hectic fever recurs.

Mr. B's wife finally solves the riddle. She notices that her husband's bouts of fever seem to coincide with his taking some antihypertensive pills (α-methyldopa) which had been prescribed to him previously. The treatment had been interrupted while he was at the hospital. She goes through a course of 'experiments' and 'counterexperiments', and when she is pretty sure the pills can be incriminated, she offers the diagnosis to the medical profession. Doctors can only confirm that it is possible for any drug to be responsible for a fever, and that a few such cases have been reported with α-methyldopa. The drug is withdrawn. The febrile syndrome disappears. The patient's temperature is thereafter normal. (See *Le Concours Médical* **104** (1982), 575– 576.)

Case history *B* points to the fact that the search for causes can be to some extent pre-theoretical, that it may include invention and dispense altogether with laws, either universal or statistical (likelihood functions). Mr. B's wife discovers the cause of her husband's discomfort through trial and error, without having much of a preconceived idea of what qualifies as a possible cause in the circumstances. The medical staff goes through the standard probabilistic procedure and misses the diagnosis.

Fact: Mr. B's oscillating fever, of two weeks' duration on admission to hospital. Wanted: the cause of the fever (etiological diagnosis).

The probabilities to be considered in dealing with a patient with prolonged (more than two weeks) fever of unknown origin (FUO) are discussed in all textbooks ([9], Ch. 16). Physicians are reminded that they should look for (the order is by decreasing frequencies):

H_1 — infection, either focal (e.g., Charcot's biliary fever, infection of the urinary tract, bacterial endocarditis, etc.) or general (tuberculosis, brucellosis, malaria, salmonella infection, rat-bite fever, relapsing fever — due to spirochetal organisms, histoplasmosis, etc.);

H_2 — neoplasm, particularly: carcinoma of the stomach or pancreas, Hodgkin's disease, acute leukemia (or other hematopoietic disease);

H_3 — disseminated lupus erythematosus or some other collagen disease, or a vascular disease (arteritis: Horton's disease, or migratory thrombophlebitis), or rheumatoid arthritis;

H_4 — endocrine disorder;

H_5 — malingering (counterfeited fever);

H_6 — other, miscellaneous, e.g., drug fever, familial mediterranean fever, etc.

The medical team proceeds quite rationally along the lines of the standard probabilistic diagnosis. They start from a set of etiological hypotheses, complete given the current state of knowledge, exhaustive (cf. H_6: 'other'), and more or less mutually exclusive (above list). The initial conjecture is a probability distribution over the set of hypotheses. A priori, the most probable hypothesis is H_1: infection. An estimate of $P(H_i/\text{fever})$ for each i ($i=1$ to $i=6$) can be gathered from frequencies. Out of 100 people with a fever, say 80 have some kind of infection, 10 have cancer, 5 have a collagen disease, 1 has an endocrine dysfunction, 1 is a malingerer, 3 have something else. Thus, a priori, $P(H_1/\text{fever}) = 0.8$.

Then doctors look for pertinent information. They collect two kinds of signs (or evidence, E):

(1) 'negative' signs, e.g., the patient did not travel abroad, he does not remember being bitten by a rat, he has no history of allergy, no anomalies on X-rays . . . . These are signs allowing (provided their sensitivity is high) one to 'eliminate' a hypothesis, in fact make it less probable (than a priori). Thus the probability that the patient has malaria, given the facts that he has an oscillating fever, he did not travel abroad, he does not live in the vicinity

of an international airport, and blood samples do not demonstrate any plasmodia, is lower than the probability that the patient has malaria, given the mere fact that he has an oscillating fever. Note one 'treacherous' sign: the patient has no history of allergy.

(2) 'positive' signs, e.g., an enlarged prostate, an elevated leukocyte count, a possibly infected tooth. . . . These are signs allowing (provided their specificity is high enough) one to 'confirm' a hypothesis, in fact make it more probable (than a priori). Note that here several traits are highly suggestive of infection: abrupt onset, high fever with chills, elevated PN count. . . . One trait is almost pathognomonic: staphylococcus aureus has been demonstrated in the blood. Thus the probability that the patient has some kind of infection, given the facts that he has a fever, his PN count is elevated, etc., and staphylococcus aureus has been demonstrated in the blood, is much higher than the probability that the patient has an infection, given the mere fact that he has a fever.

The physicians' conclusion is correct (rational) — but it is false. They think that, given the available evidence, a posteriori, by far the most probable hypothesis is that of infection. They know that there is a doubt, because (1) the leading symptom (fever) has vanished, as well as the most significant sign (the presence of staphylococcus aureus in the blood) — retrospective diagnosis is precarious; (2) crucial information may be missing.

Now imagine that one of the physicians was particularly interested in drug fever. He insisted on asking: "what drugs did you take?", and he did collect the information that the patient had been taking α-methyldopa (among other drugs, like aspirin, laxatives, antibiotics, etc.). One may doubt whether the standard (Bayesian) inference could by itself have led him to the true diagnosis. Schematically the inference might go this way (the figures are speculative):

— a priori probabilities: of infection, $P(H_1/\text{fever}) = 0.8$; of drug fever, $P(H_6/\text{fever}) = 0.03$;

— factual information and likelihoods: E_1: elevated PN count, and presence of staphylococcus aureus, and possibly infected tooth, $P(\text{fever}/H_1) = 0.8$, $P(\text{fever}$ and $E_1/H_1) = 0.9$, $P(\text{fever}$ and not $E_1/$ not $H_1) = 0.7$, etc.; E_6: drug taken, and no known history of allergy, $P(\text{fever}/ H_6) = 0.00001$, $P(\text{fever}$ and $E_6/ H_6) = 0.000001$, $P(\text{fever}$ and not $E_6/$ not $H_6) \simeq P(\text{fever}$ and $E_6/$ not $H_6)$, etc.

— quite possibly, when figuring out the a posteriori probabilities, one would find that: $P(H_1/ \text{fever}$ and E_1 and $E_6) > P(H_6/ \text{fever}$ and E_1 and $E_6)$.

After the true hypothesis has been identified, of course those concerned reckon that they should have thought of it, but they should have thought of a thousand other things as well.

Mr. B's wife proceeds empirically. She observes that her husband had a hectic fever at home, that he became apyretic after six days in the hospital, that the fever returned when he was home again. What made the difference? It must be a circumstance present in her husband's life at home, and absent from his life at the hospital. There are many of those. Timing is the clue. She suspects that the taking of antihypertensive pills may be the circumstance, because the bouts of fever seem to follow his taking the medication. She has no pathogenetic theory. She does not know that α-methyldopa may induce febrile reactions. She does not even know what α-methyldopa is. To confirm her suspicion, she resorts to trial and countertrial, i.e., experimental method in C. Bernard's sense (1865). Call C the taking of the pill, E the fever. She observes that when C occurs, then E occurs; when not C, then not E. A coincidence is not excluded. She does not try a great many times. She goes to the learned physicians for ratification. They know pathogenetic theories, whereas her approach has been rough and naive. Nevertheless her explanation is the explanation, they missed the diagnosis and she got *the* cause.

Physicians recognize the cogency of Ms. B's observation straight away, as though all the signs previously interpreted from the perspective of infection suddenly assumed a different and obvious meaning in the light of the drug hypothesis. At the next staff meeting, Mr. B's case is turned into an educational precept. Here is roughly what the attending students hear:

(1) General principle: "no drug can a priori be considered innocent in a fever case". This is not very informative. It merely says that the taking of a drug is one of the hypotheses to be contemplated in the etiological diagnosis of a fever, i.e., P(drug taken/ fever) $\neq 0$, even though P(fever/ drug taken) is very low.

(2) When collecting indices: One should not allow oneself to be perplexed or disoriented by such signs as: sudden onset, chills, elevated PN count, oscillating pattern of fever; none of these is specific to infection; even the presence of germs in the blood is not pathognomonic of an infectious etiology! One should think of always asking the patient what drugs he has been taking; if he has taken any, one should reflect that iatrogenic drug-induced fevers may have a sudden onset, may be intermittent, are often accompanied by an elevated leukocyte count, and finally that the presence of staphylococcus aureus in the blood may be an *effect* of high fever (bacteriemia induced by a febrile reaction). Thus, all the signs are reinterpreted.

(3) If it is strongly suspected that the drug is responsible, it should be discontinued. If the fever recedes, the drug can probably be incriminated.

A high probability does not amount to certainty. If higher probability is desired, taking of the drug should be resumed. But going that far may be unethical. "Any questions of drug fever can be resolved rapidly by discontinuing all medication. The diagnosis can be further substantiated by giving a test dose of the drug after fever has subsided, but this may result in a very unpleasant or even dangerous reaction" ([8], I, 16).

In her ingenuity, Ms. B did precisely what the learned physicians might not have allowed themselves to do. She not only identified the cause in the absence of any theoretical background, she got the beginning of a proof. The physicians' contribution is then to recognize that her finding is indeed compatible with medical knowledge.

It is well known that Bayesian types of inference may for a while give high probabilities to false hypotheses (when significant information is missing), and that diagnostic procedures, which cannot last indefinitely, are bound to end with a decision, even in the absence of 'conclusive' evidence. The decision to admit of an elusive infectious etiology and send the patient home was certainly wise in case B. This much can be said in favor of a probabilistic Bayesian analysis of the causal inquiry, that it gives a competent account of the pragmatic aspect of causal determinations. Such a pragmatic approach was here contrasted with a more empirical approach.

3. PATTERN RECOGNITION VS. THEORETICAL BIASES: CASE HISTORY C

Ms. C was an assistant bookkeeper in an industrial firm (chemicals) until (in 1975, she was 54 years old), due to failing memory and loss of intellectual efficiency, she had to be transferred to a lower grade position. Later, her mental condition became so defective that she stopped working and was hospitalized in neurology.

On admission she was found to be coherent, with fair vigilance and attention, even some critical acumen, no aphasia, apraxia, or agnosia. Her short term memory appeared to be seriously impaired, and the general intellectual performance was poor, given her training. The examination of the motor system showed increased tendon reflexes (evoking a pyramidal disorder).

The EEG was subnormal: no focal or localized slow wave or seizure activity, poorly structured alpha rhythms. Radiologic examinations were normal, in particular, there were no signs of low pressure hydrocephalus.

The diagnosis was: progressive dementia in its early stage. The patient was released from the hospital and sent to a convalescent home because she had fractured a bone (astragalus).

Two years later Ms. C was hospitalized again. The motor syndrome had aggravated: straight pyramidal disorders with bilateral Babinski's sign, rigidity and a tendency towards retropulsion, instability of posture and inability to walk. Sphincteric incontinence had set in. The patient was apathetic, apragmatic, inert, often somnolent. Although she lacked insight, she did not seem to be indifferent to the changes in herself, and her mood was depressed. When stimulated enough, she would answer questions and stand some tests; she could talk, calculate, identify objects, carry out some simple tasks. Her judgment was massively impaired, her general comprehension was limited, her attention was weak and she was extremely distractible; her memory was very defective, particularly for recent events. The radiographic examinations (scannography and Pantopaque ventriculography) showed an enlargement of the lateral and third ventricles, suggesting a stenosis of the aqueduct of Sylvius; they also showed a bizarre pattern of clustered radiolucency, interpreted as ischemic lesions. A tentative surgical treatment (ventriculocisternotomy) brought a transient improvement: Ms. C looked less apathetic, did better in rehabilitation. Soon however, a dramatic deterioration of behavior took place, ending with stupor, right hemiparesia, coma, infection and death (June, 1980).

Pathologic findings: The macroscopic examination of the brain showed: (1) a sowing of 'cysts' (0.5 to 10 mm in diameter) in the middle brain (mesencephalon) and thalamus; (2) a major symmetric enlargement of the lateral and third ventricles, connected with a collapse of the aqueduct, constricted by the cysts.

The histologist in charge of the microscopic examination of the sections diagnosed a lacunar state, in fact multiple infarction in the vascular territory of the thalamo-geniculate artery (Percheron artery).

Such a diagnosis was questioned because the patient had neither hypertension, nor a particularly striking degree of arteriosclerosis, nor any of the known cardiovascular risk factors.

Intrigued by the very peculiar cystic pattern, the histologist consulted with international experts. Lacunae or not lacunae?

The 'American school' argued that extensive cystic lesions had never been described as 'lacunae', that the use of the term was certainly improper in this case, and that it had not been proved that the cysts did not have an origin other than vascular (e.g., congenital).

The 'British school' confirmed that the cystic cavities were indeed common lacunae, located in their ordinary sites of election.

It was then realized that the two schools had entirely different pathogenetic theories of lacunae, although both claimed that lacunae are the result
of hypertension combined with atherosclerosis.

'American' theory: lacunae are limited foci of cerebral necrosis with ir-
regular borders, often pale but sometimes orange-yellow, due to thrombotic
occlusion of small terminal penetrating arteries running to the middle cerebral
area. Arteriolar occlusion results in infarction (possibly hemorrhagic) of the
down-stream territory. The clean-up of the territory by the macrophages
results in lacunar disintegration, surrounded by a zone of astrocytic reactive
gliosis. Hyalinized blood vessels and discrete sequelae of hemorrhagic lesions
(hemosiderin pigment) are frequently noted.

'British' theory: lacunae are pale cystic cavities, with a thin smooth wall
made of a simple squamous epithelium. In small lacunae a blood vessel may
often be seen floating in the cavity. Atherosclerosis and hypertension result
in crooked pulsating arteries, inducing an enlargement of the perivascular
spaces of Virchow-Robin. (Dilatation of the perivascular spaces was first
observed by Durand-Fardel, 1842.) Lacunae are precisely the perivascular
dilated spaces.

The 'French school' reckoned that there probably were at least two kinds
of lacunae. The pale cystic cavities observed in the case of Ms. C resembled
the lacunae of the British school, even though the British pathogenetic theory
was doubtful. The question of whether the course of Ms. C's dementia might
have been hindered could not be answered. (See *Hôpital Henri Mondor*,
94010 Créteil, France. Neurology, Neurosurgery and Histology Departments.
Pathology: Case nb. A4897. See also [11]. The case is cited with the kind
permission of J. Poirier, M. D., Head, Department of Histology, Embryology,
and Cytogenetics.)

Case history C is paradigmatic of the extent to which positive (*What?*)
and etiological (*Why?*) diagnoses are intermixed. Positive diagnoses, when
contrasted with etiological diagnoses, are supposed to yield the nature of
the ailment, not its cause. Yet naming a pattern of behavior ('progressive
dementia') or a lesion ('lacunae') amounts already to designating a sort of
cause (formal or material cause, in Aristotle's sense?). At the same time such
observations call for further causal inquiry: What led to the dementia? What
produced the lacunae (efficient cause, in Aristotle's sense)? Artificial intel-
ligence research has thrown some light on the play of categorical versus
probabilistic components in medical reasoning strategies. Pattern recognition techniques are used to extract significant features from raw data (for
example, to discriminate between normal and abnormal ECG, categorize

abnormalities, and label them, e.g., 'left ventricular hypertrophy'). Such a labeling, however, rests on a causal inference from observation (of the wave pattern) to recognition (of a state of the heart). In fact, some of the currently available automated (ECG) pattern interpretation programs have been conceived within a Bayesian probabilistic framework. Conversely, the ideal pathogenetic theory of lacunae would be a typical description of how lacunae spring up and are shaped in the brain. The choice (possibly involving Bayesian and decision theory procedures) of a causal event initiating the course of things ending with a lacuna would only be a matter of pragmatic concerns (e.g., from what stage could Ms. C's mental deterioration have been prevented?). Finally, neither the causal history which is at the bottom of the pathogenetic theory, nor the recognition of the lacunary pattern, are entirely amenable to a probabilistic analysis.

From age 54 to 59, Ms. C deteriorated continuously and became senile, a most unusual fact which requires explanation. In accordance with the tradition of anatomo-clinical methods, her condition was imputed to anatomical lesions (material cause). The (ischemic?) nature of those lesions was not clear to clinicians, although Ms. C underwent brain surgery on the (causal, probabilistic) hypothesis of low pressure hydrocephalus. (Low pressure hydrocephalus is one of the known possible causes of early dementia.) Macroscopic pathologic findings assessed the (positive? etiological?) diagnosis: there was a combination of holes in the brain (positive finding) and compression of the aqueduct (causal finding: it is known that a lesion in the vicinity of the aqueduct may result in compression, that compression of the aqueduct results in hydrocephalus, and that hydrocephalus results in mental impairment). Microscopic examination of brain sections yielded the (positive? causal?) diagnosis of the holes: these were lacunae. Why did Ms. C develop such extensive lacunae? It is commonly admitted that lacunae are the result of hypertension and atherosclerosis (these are risk factors, i.e., causes in the probabilistic sense). Ms. C had neither of those.

The histologist reports that when he examined the brain sections, he had no doubt or hesitation: he immediately recognized lacunae. As far as he can tell, that was not a (probabilistic, or other) inference, but an unanalysed Gestalt identification. He compares it to dermatologic diagnosis 'at first sight'. Computers indeed use sophisticated mathematical and statistical techniques to detect simple patterns. But research in pattern recognition developed from engineering and aimed at achieving accurate performance, not at approximating human perceptual behavior. The histologist claims that − no matter how visual information was processed in his brain − he felt

immediately certain that the holes were lacunae, and that he resolved to
consult with international experts not because he doubted his (say, positive)
diagnosis, but because no (causal) explanation was available and the lacunary
pattern was extraordinarily extensive.

The comments of the consultant experts suggest that, contrary to what the
histologist tends to think (namely, that these are indeed lacunae, whatever
doubt there may be as to their cause), pattern recognition (positive diagnosis)
is probabilistic after all, and unconsciously dependent on etiological knowledge. The American consultant does not see lacunae, whereas the British
expert recognizes lacunae straight away. It is then realized that the concept
of 'lacunae' is heterogeneous, referring to two pathogenetic theories (the
supporters of which ignore each other). They can be outlined this way:

In either case the primary cause is believed to be a combination of hypertension and atherosclerosis. Americans think that hypertension results in
thrombotic occlusion of small atherosclerotic arteries, which results in
infarction of arteriolar territories, which results in holes after the softened
areas have been cleaned up by the macrophages. British neuropathologists
hold that the hypertensive pulse makes fibrous arteries become crooked,
which breaks the perivascular spaces more open in places, which results in
the lacunar pattern.

Interestingly enough, the description of lacunae in neuropathology textbooks is ambiguous, and compatible with either opinion. Does it imply that
(in most cases, at least) the recognition of cerebral lacunae is independent
of etiological knowledge, after all? Grossly, it may be. Traditionally, lacunae
are little holes, less than 10 mm in diameter. Precisely, it can't be. Should
the British hypothesis be true, the walls of the holes would be lined by
an epithelium. Should the American hypothesis be true, they would not.
Epithelia are easy things to identify for histologists. In the case of Ms. C the
walls of the holes were lined by an epithelium.

Given the criterion, it is clear why some rejected and others welcomed
the lacunar hypothesis in case C. No-one mentioned such a criterion, however. It is as though the presence or absence of an epithelium had been an

unconscious, unanalyzed element of the familiar visual pattern, and had not been theorized about.

The French histologist reckoned that there might be at least two, possibly three types of lacunae, depending on different etiologies: (1) small infarcts, (2) small hemorrhages, (3) dilated perivascular spaces. Distinctive characteristic features of each could be described. Whether the various types could be traced back to a common cause (such as hypertension) was an open question. The American concept coincided with the first type and the British concept with the third. The probabilistic character of the (positive) diagnosis was related to the concept's residual vagueness. Obviously lacunae in American brains could not have been all of the first type, while they were of the third type in English brains. It is well known that researchers have a tendency to neglect data which do not fit their theories. One finds it hard to believe that they cannot see them. Perhaps the other type could not be seen as such because it could not be explained, or perhaps it could not be categorized. Note that, if after distinctive types have been particularized, a decision is made to save the word 'lacunae' for the first (or third) type only, and use other names for other types, such a decision does not involve any probabilistic inference.

Now one thing is to hypothesize about lacunae in general, another thing is to explain why Ms. C deteriorated. The man in charge of the post mortem examination is supposed to lay out the lesions which induced the clinical condition and ultimately led to death. He typically reconstructs some sort of causal history of the case, the raw material of which is a mixture of clinical data and pathologic findings; causal analysis is used to single out an overall causal sequence from the manifold of actual facts. Ideally such a reconstruction should be probabilistic only to the extent that significant (with respect to the current state of knowledge) causal factors are inferred from the observation of their effects, and such factors are not believed to have completely determined subsequent events. For example, there is little doubt that the development of lacunae, together with the compression of the aqueduct it apparently entailed, caused Ms. C's progressive dementia (among the known possible causes, this is one, and the most probable in the circumstances). However, as lacunae are commonly found in the brains of people who show no mental impairment, and there is no evidence of a definite relation between, say, the loss of cerebral substance and the severity of the dementia, and since compression of the aqueduct by the lacunae has not been demonstrated with certainty, the alleged cause cannot be considered to have automatically produced the clinical syndrome. On the other hand, the clinical syndrome is a fact, as are the loss of cerebral substance and the localization of the lacunae in the territory of the thalamo-geniculate artery. Where there are major gaps in the factual basis, no amount of causal analysis can bridge them. The knowledge of typical causal sequences can only help discern data the importance of which might have been otherwise overlooked. No effect can be imputed to a cause which was not known to be present. If it is established that Ms. C had no hypertension, or even if it is not positively established that she had hypertensive episodes, there is no way that hypertension can be made the primary cause of her disease, even though available pathogenetic theories unanimously incriminate hypertension as the initial disorder resulting in lacunae. While causal links are bound to be hypothetical to some extent, i.e., probabilistic, hypothetical facts inevitably undermine the most sagacious etiological reconstruction.

4. INVESTIGATION OF A NEW MORBID ENTITY: CASE HISTORY *D*

Ms. D was born in 1927. She lived over 40 years without any medically significant problem except a treated tuberculosis (1955). Around 1973 her physical aspect started changing: thickened skin, hirsutism, sweating, cyanosis of the fingers and nails (acrocyanosis); she also probably developed slight neurological troubles that went unnoticed at the time. Then, between 1977 and 1981, she was hospitalized repeatedly, either in gastroenterology or in neurology, for a variety of seemingly heterogeneous disorders.

1977, March: acute hydroelectrolytic diarrhea. No specific cause was found. The diarrhea receded under symptomatic treatment. Routine investigations displayed two anomalies: (1) a slight renal insufficiency, attributed to hyperuricemia (asymptomatic gout), for which she was prescribed an uricosuric drug (allopurinol); (2) acrocyanosis and Raynaud's phenomenon, with no other evidence of an altered immune response than a little IgA-λ chain 'spike'.

1977, May-June: splenomegaly, general lymph node enlargement. She was splenectomized. The spleen weighed 1 kg. The histology showed a nonspecific congestive and inflammatory aspect. Samples from the lymph nodes and liver were also congestive.

1977, October: elevated platelet count (thrombocythemis). A biopsy of the bone marrow was normal. A little IgA-λ chain spike was detected in the blood. The patient was prescribed aspirin.

1978, May: neurological disorders. Clinical findings: (1) motor disturbances of the lower limbs (tendon reflexes were abolished, there was difficulty walking downstairs, and tenderness of the muscles); (2) Argyll Robertson pupil (anisocoria); (3) papilledema. Laboratory findings: (1) abnormal oral glucose tolerance test; (2) electromyogram: partial denervation of muscle, low conduction velocity of nerve (22 m/s); (3) neuromuscular biopsy: non-specific denervation, negative immunofluorescence test. Conclusion (confirmed by a consultant expert): diabetic neuropathy. The patient was sent home with a treatment for her presumed diabetes.

1979, March: abdominal pain, liver and lymph node enlargement. The patient was thoroughly explored. Various abnormalities were found. Clinical examination: (1) alterations in the integument (thick skin, telangectasias, hirsutism, excessive sweating); (2) slight tachycardia (otherwise auscultation of the heart was normal). Instrumental and laboratory examinations: urine: slight proteinuria (0.25 g/24h); blood: elevated white cell and platelet (1 million) counts; creatinine 18.5 mg/1; glycemia and uricemia were normal; immunologic reactions: (1) unchanged IgA-λ chain small spike, (2) antinuclear antibodies (positive at 1/10 dilution); liver: normal liver function tests; scintiscan, EMI-scan: homogeneous hepatomegaly; arteries: catheterization of all major arteries was performed, in particular of both carotid arteries; arteriography demonstrated several narrow stenoses; during arteriography of the right int. carotid, the patient had recurring episodes of transient monocular blindness (treated with aspirin).

At this point a diagnosis was suggested. This might be a case of *plasma cell dyscrasia with polyneuropathy and endocrine disorder*, a rare syndrome, first described by Shimpo (1968) under the label *solitary plasmocytoma with polyneuritis and endocrine disturbances* in the *Japanese Journal of Clinical Medicine* **26**, 2444. Thus far, some 30 such cases had been described, 25 of which were found in Japan.

More investigations were performed in order to substantiate the diagnosis, in particular: endocrine evaluation: no hormonal abnormality was detected (note that signs of diabetes mellitus were now absent); skeletal radiological survey: possibly a focal sclerotic lesion within the left humerus (?); biopsy of both temporal arteries: atherosclerosis.

It is reported in the file that Ms. D now felt "depressed and weary". She was not seen at the hospital for two years, except for episodic consultations. At one point she was prescribed chemotherapy (cyclophosphamide) without any apparent effect. (Did she take the drug?)

1981, December: acute diarrhea, abdominal pain. Hyponatremia was diagnosed. To correct this abnormality the patient was administered intravenously an hypertonic saline solution. The result was overcorrection of the hyponatremia, and massive edema. Diuretics were given to induce sodium depletion. A pericardial friction rub appeared. Pericardiocentesis was instituted to reduce the pericardial effusion (0.85 liter of serohemorrhagic fluid was withdrawn), and a biopsy of the pericardium was performed. The heart stopped repeatedly and eventually did not start again.

Pathologic findings: Widespread enlargement of the lymph nodes, hepatomegaly (2.6 kg); discrete atrophy of left kidney; moderate atheroma of the aorta and its branches (fatty streaks without calcifications); discrete pulmonary edema, discrete pleural effusion, pericardial effusion without myocardial damage.

According to the pathologist, none of the observed lesions explained the disease. The pathologist collected whatever information he could find on the syndrome in the literature.

The P.O.E.M.S. syndrome: The syndrome was first described as a unique entity by Shimpo (1968). Thirty-nine cases were reported by 1979, most of them in Japan (and in Japanese); cf. Takatsuki *et al.* (1976), 'Plasma Cell Dyscrasia With Polyneuropathy and Endocrine Disorder', reviewed in *Excerpta Medica*. A report of two more cases and a review of the literature available in English were given by Bardwick *et al.*, in [1], 'Plasma Cell Dyscrasia With Polyneuropathy, Organomegaly, Endocrinopathy, M-Protein and Skin Changes: The P.O.E.M.S. Syndrome', in *Medicine* **59**, 311–322.

Here is a brief outline of current knowledge. The P.O.E.M.S. syndrome is a multisystem disorder, the pathogenesis of which remains totally obscure. The course of the disease is chronic. After a prolonged course the patients may die from complications of the neuropathy. The syndrome includes five main features:

(1) *Polyneuropathy* (resembling a Guillain-Barré syndrome, which is thought to be an immunologically mediated inflammation of the dura mater, possibly induced by a virus). Two sets of traits dominate the picture: Peripheral sensori-motor neuropathy with extremity paresthesias or dysesthesias, absent tendon reflexes, muscle weakness; then glove and stocking anesthesia, arm and leg palsy accompanied by muscle wasting. Diminished visual acuity, papilledema, elevated intracranial pressure, increased cerebrospinal fluid protein (in fact, generalized edema, including ascites and episodes of anasarca).

(2) *Organomegaly*: enlargement of liver, spleen, lymph nodes. Liver function tests are normal. Biopsies are normal, apart from nonspecific hyperplasia. No amyloidosis, no hemochromatosis, no porphyria.

(3) *Endocrinopathy*: failure of multiple endocrine organs, including amenorrhea in women, impotence and gynecomastia in men. X-rays of the sella turcica are normal. There is hyperprolactinemia, with elevated gonadotropins, suggesting primary gonadal failure. Glucose intolerance (diabetes). Sometimes hypothyroidism or adrenal insufficiency. (It is hypothesized that the hypothalamic dysfunction may be due to increased intracranial pressure.)

(4) *M-protein*: in fact, a bone and bone-marrow disorder, with: osteosclerosis, including: (1) focal areas of bone sclerosis (with or without bone destruction), (2) fluffy spiculated areas of osseous proliferation (particularly in sites of ligamentous attachments). Plasma cells may be found in the bone lesions. In one case radiation therapy of a solitary osteosclerotic lesion brought a dramatic improvement in the patient's clinical status (normalization lasted two years; when the symptoms returned, a new bone lesion was discovered). Plasma cells are slightly more numerous than normal in the bone marrow (although quite scarce). The disease has been called 'solitary plasmocytoma', 'plasma cell dyscrasia', 'osteosclerotic plasmocytoma', reflecting the belief that the primary lesion might be a plasma cell tumor. Transient hypergammaglobulinemia; immunoglobulin M-component containing only λ light chains.

(5) *Skin changes*: hyperpigmented, indurated skin, with telangectasias; increased hair growth (hypertrichosis), hyperhydrosis. Cutaneous biopsies show dermal edema. In some cases vascular lesions have been reported. Characteristic features of scleroderma, polymyositis, or systemic lupus erythematosus are absent. Some features are analogous to toxic manifestations induced by chronic exposure to trichlorethylene.

The Japanese (tentative) pathogenetic hypothesis is formulated with prudence by Bardwick *et al.* [1]: "It is hard to avoid implicating some product of the plasma cells (antibodies) as a pathogenic factor, because of the occasional dramatic symptomatic improvement following local treatment of the tumor." (See *Hôpital Henri Mondor*, 94010 Créteil, France. Neurology and Histology Departments, Drs C. Meyrignac and J. Poirier. Pathology: Case nb. A5425, Dr F. Lange).

Induction by elimination presupposes that there are hypotheses to be confirmed or disconfirmed. It does not create any new hypotheses (just as neo-Darwinian selection does not create any new genes). Case history *D* shows how, when a causal hypothesis has not been clearly formulated yet, the standard procedure of probabilistic (Bayesian) diagnosis is apt to err and, to the extent that it is effective while beating the air, to be hazardous for the patient.

Fact: a rich and varied symptomatology was exhibited by Ms. D. Wanted: the disease entity accounting for the symptoms. Note that there is no clear explanation of why it is assumed that there was *one* illness underlying the various symptoms, rather than several different pathologies.

As Ms. D was examined at the hospital repeatedly, physicians went through a process of screening all the possible diagnoses, more or less in the Bayesian mode. Each time a new set of symptoms appeared, the previous diagnosis was reformed, some other hypothesis appeared more probable. The new hypothesis was investigated, and whenever confirmed by laboratory or other findings, it led to tentative therapy, until the clinical picture took a different turn, and outmoded the leading hypothesis again, and so on. From an exhaustive list of hypotheses you cannot get an unknown diagnosis through a correct inference, you get: 'other, miscellaneous'. By the time that you have ended up with 'other', and feel that this is a puzzle you cannot make sense of in the present state of your information, the patient may already have undergone quite a number of diagnostic tests and therapeutic attempts, of which it is unclear what the returns are for her. In the case of Ms. D, here are some of the tracks that were followed and abandoned:

(1) Blood tests suggest a slight renal insufficiency. What is it due to? Blood tests also show a mild hyperuricemia (gout). Renal insufficiency may be due to hyperuricemia. Therefore allopurinol is prescribed to normalize uricemia.

(2) Acrocyanosis and Raynaud's phenomenon are conspicuous in the patient. Those may indicate a connective tissue disease. In that case immunologic abnormalities should be evidenced. (Very little was found on that track, which led to no treatment, but to iterative exploratory tests).

(3) The spleen is enlarged. Splenomegaly may be related to Hodgkin's or other malignant disease. Radical treatment: the spleen is removed. (Remember that, after the splenectomy, Ms. D's liver started enlarging).

(4) An elevated platelet count suggests hematopoietic cancer. Hence sophisticated investigations (biopsy of the bone marrow, etc.), if only symptomatic treatment (aspirin).

(5) Argyll-Robertson pupil brings up a suspicion of syphilis, which is not probed urgently because such a diagnosis is hardly the fashion now.

(6) The peripheral neuropathy may be (is, indeed, by a consultant expert) ascribed to diabetes, given an abnormal oral glucose tolerance test. Standard treatment of the presumed diabetes is prescribed. (It is doubtful whether Ms. D always followed the treatments prescribed to her. Note that a year later biological signs of diabetes had vanished anyhow). Etc.

There came a clue, at last. Someone mentioned the 'Japanese disease', someone looked into the literature. Many a trait of Ms. D's case fitted into the P.O.E.M.S. syndrome: peripheral neuropathy, spleen, liver and lymph node enlargement, glucose intolerance, skin changes, immunoglobulin λ light chains. However the nature of the disease was still hypothetical; so far there had been very few cases reported. The overall pattern was elusive, it had to be elucidated.

From then on Ms. D became an interesting specimen through which the paradigmatic entity was to be deciphered. Her helplessness was taken advantage of to get information. She lost her spleen, various pieces of liver, lymph nodes, arteries, muscle, nerve . . . discounting the numerous samples of blood, cerebrospinal fluid. . . . The investigation was fruitless for her, it was even harmful (she was disheartened; she died an untimely, iatrogenic death, in the course of a disease described as chronic). Incidentally, American patients in whom the same syndrome was identified were scrutinized with an even more invasive technicality.

This is not to say that the consequences of conducting clinical research are devastating. When she went to the hospital Ms. D was looking for help, and the best help that can be offered to someone with an enigmatic disease, besides words of comfort, is to set about studying the disease. The point is this. Weighting the risk and cost of diagnostic tests against their anticipated informative import, and deciding whether it is worth getting a higher degree of confirmation for a given causal hypothesis at the cost of further inquiry, is already a delicate matter within the standard set-up of choosing one among a set of preformed hypotheses. It is all the more difficult, nay impossible, as the causal concept has to be invented, and one is unaware of what kind of investigations are relevant. Doctors are familiar with the idea that some diagnoses had better stay unexplored; for example, in most presumed viral infections, the identification of the virus responsible for the infection is not pursued, because the time and energy consumed in laboratory tests would not be compensated by any therapeutic or intellectual advantage. That sort of implicit calculation makes little sense, or becomes subjective to the extreme, when as in Ms. D's case the disease entity is ill-defined. How is one to guess what the strategic utility of some or other test will be? Yet, however desirable it may be to break through with a meaningful discovery, surely not everything can be tackled on one patient. Given the variety of possible diagnostic explorations now available, judicious choices should be made. On what rationale? Apparently, not only the probabilistic (Bayesian) causal analysis, but the whole (neo-Bayesian) rational decision process is held in check here.

Observation and analogy are the researcher's guides when exploring an illness hitherto unknown. From the first few case histories collected a schematic disease pattern is sifted out, on the basis of the main common clinical and biological findings. Then through a screening of all the known disease patterns analogous or partially analogous to the present one, tentative causal hypotheses are tried. For the unity of the disease concept is thought to come from etiology. Contrary to what happened in the standard well-buoyed situation (above), there are no rules or systematic strategies to be applied in that search. Here is roughly the kind of speculation that went on recently around the P.O.E.M.S. syndrome [1] :

(1) Some aspects of the syndrome are suggestive of multiple myeloma, a neoplasm of plasma cells, with bony lesions due to the development of tumor tissue in the bone marrow, and overproduction of an abnormal immunoglobulin component. P.O.E.M.S. might be a 'mild' form of such a cancer (plasma cell dyscrasia). In fact, radiotherapy of the bone lesions was in some cases effective in relieving the symptoms. But an association of osteosclerotic bone lesions and the typical clinical features of the P.O.E.M.S. syndrome have been observed in the absence of any immunoglobulin abnormalities or plasmocytosis.

(2) The neuropathy evidenced in P.O.E.M.S. patients may be ascribed to diabetes mellitus: many P.O.E.M.S. patients have exhibited some degree of glucose intolerance at some point, many diabetics develop a peripheral sensorimotor neuropathy. But the neuropathy is far more constant and massive here than that which the fitful hyperglycemia would allow. Another explanation for the neuropathy would be that an abnormal immunoglobulin, or a fragment thereof, is produced by the plasma cells and deposited on the membranes around the nerves, as in the Guillain-Barré Syndrome. But immunofluorescence has not demonstrated anything of the sort yet.

(3) Some of the observed traits evoke scleroderma (tight, hyperpigmented skin), or polymyositis (tender muscles), or systemic lupus erythematosus (impaired renal function), or periarteritis nodosa (lesions of arteries). Hence P.O.E.M.S. might belong in the group of so-called auto-immune diseases. But, so far, no auto-antibodies have been detected.

(4) The endocrine dysfunction is rather enigmatic. Gonadal failure may induce hyperprolactinemia, but what induced the gonadal failure? Anasarca (including increased intracranial pressure) may explain hypothalamic dysfunction, but what is the cause of the anasarca?

(5) The chronic intoxication hypothesis is seductive. Exposure to trichlorethylene occasions toxic manifestations quite similar to some of the P.O.E.M.S.

manifestations. But no chronic intoxication has been demonstrated in the recorded patients. . . .

None of the putative etiologies has been convincingly evidenced yet. Japanesé authors favor the first hypothesis. The syndrome is currently being studied further as new cases have been detected. In the early stage of its investigation it appears as a multisystem disorder the cause of which cannot be invented through any standard probabilistic (Bayesian) inference, either human or computerized, because only standard diagnoses can be found that way, new concepts are formed some other way.

Finally, the reader may have perceived that in this review, the notion of a cause and the methodology of identifying causes have not been separated. This writer thinks that they are not separable [4] .

University of Paris XII, and
C.N.R.S., Paris, France

BIBLIOGRAPHY

[1] Bardwick, P. A. *et al.*: 1980, 'Plasma Cell Dyscrasia with Polyneuropathy, Organ-omegaly, Endocrinopathy, M-Protein, and Skin Changes: the P.O.E.M.S. Syndrome', *Medicine* **59** (4), 311–322.

[2] Bernard, C.: 1865, *Introduction à l'étude de la médecine expérimentale*, Paris.

[3] Bernard, C.: 1947, *Principes de médecine expérimentale*, Presses Universitaires de France, Paris, posthumous.

[4] Fagot, A.: 1980, 'Probabilities and Causes: On Life Tables, Causes of Death, and Etiological Diagnoses', in J. Hintikka *et al.* (eds.), *Pisa Conference Proceedings*, Vol. 2, *Probabilistic Thinking, Thermodynamics, and the Interaction of the History and Philosophy of Science*, D. Reidel Publishing Co., Dordrecht, Holland.

[5] Fagot, A.: 1982, 'Calcul des chances et diagnostic médical', *Traverses* **24**, 85–102.

[6] Fagot, A. (ed.): 1982, *Médecine et probabilités*, Université de Paris XII and Didier-érudition, Paris.

[7] Fetzer, J. H. (ed.): 1981, 'Probabilistic Explanation', *Synthese* **48** (1 and 2).

[8] Gombault, A.: 1982, 'Une naissance constitue-t-elle un préjudice?', *Le Concours Médical* **104**, 6197–8.

[9] Harrison, T. R.: 1970, *Principles of Internal Medicine*, 6th ed., McGraw-Hill, New York.

[10] Hart, H. L. A. and A. M. Honoré: 1959, *Causation in the Law*, Oxford University Press, Oxford, England.

[11] Poirier, J. *et al.*: 1983, 'Démence thalamique, lacunes expansive du territoire thalamo-mésencéphalique paramédian, hydrocéphalie par sténose de l'aqueduc de Sylvius', *Revue Neurol.* **139** (5), 349–358.

ANNE M. FAGOT

126 ANNE M. FAGOT

[12] Soutoul, J. H.: 1983, 'Le juriste et l'embryon humain', *La Pratique Médicale* **1**, 49–51.
[13] Suppes, P.: 1970, 'A Probabilistic Theory of Causality', *Acta Philosophica Fennica*, Fasc. **24**, North Holland, Amsterdam.
[14] Trèves, R.: 1983, 'Quel est votre diagnostic?', *Le Concours Médical*, **105**–11, 1131–4.

B. CAUSAL SELECTION

STAFFAN NORELL

MODELS OF CAUSATION IN EPIDEMIOLOGY

The relationship between diseases and their causes can be approached from different starting-points. In *clinical medicine* we are confronted by a patient *P* with certain complaints, and eventually a diagnosis is reached because the clinical findings suggest that *P* has a certain disease *D*. Trying to understand what has happened, we may ask: why did our patient *P* catch the disease *D*? In *epidemiology*, on the other hand, we are dealing with the occurrence of a disease *D* in the population. (Epidemiology has been defined as the study of disease occurrence in human populations.) The purpose of epidemiological studies is often not merely to describe, but also to explain, the occurrence of *D*. Empirical data may be derived from cohort, case-control, or experimental studies where the occurrence of *D* is related to characteristics of individuals and their environment, e.g., pollution, smoking, diet, or immunization. An important aim is, of course, the *prevention* of disease occurrence by interventions against causal factors in the population. The relevant measure of disease occurrence, then will be a measure of *incidence*, such as the proportion of individuals falling ill during certain period of time (the Cumulative Incidence Rate) or the number of cases per person-year at risk (the Incidence Rate).

Descriptive epidemiological data will often suggest that there are considerable variations in the incidence of *D* from time to time (seasonal variations, changes over years or decades, secular trends), from one place to another (local variations, urban-rural differences, variations between different regions or countries) and between different groups of people (e.g., differences between socioeconomic groups). Trying to understand this, we may ask: why these variations in the incidence of *D*? Looking at the incidence of *D* in a defined population during a certain period of time, we may ask: why this occurrence of *D*?

A model describing the relationship between diseases and their causes should be *simple*, but it should also be plausible and fit *the complexity of empirical observations*. It would be simple enough to say that the disease *D* is caused by a certain factor *F*. However, in epidemiology (and I believe also in clinical medicine) this would not provide a full description of what actually happened. For one thing, a single causal factor is rarely, if ever,

129

L. Nordenfelt and B.I.B. Lindahl (eds.), Health, Disease, and Causal Explanations in Medicine, 129–135.

sufficient to explain a single case of *D*, much less to explain the occurrence of *D* in a population. For instance, while smoking may be an important factor in the causation of lung cancer, it does not seem to be sufficient since most smokers never get the disease. This suggests that there is more than one factor (smoking) involved when a smoker gets lung cancer. In fact, there may be several other factors such as heredity, radiation, and air pollution. Furthermore, a single causal factor is usually not a *necessary* condition for *D*. For instance, many people die from lung cancer who have never been exposed to tobacco smoke. Obviously the situation is different when a disease is defined by a causal factor, as is the case with tuberculosis. In most cases, however, a single causal factor is neither a necessary nor a sufficient condition for *D*.

In the causation of a disease *D* one factor may be a cause of another factor which in turn is a cause of *D*. For instance, a dietary excess of saturated fat and cholesterol may cause hyperlipidemia, which in turn causes coronary atherosclerosis and occlusion, and eventually myocardial infarction. The recognition of such *Causal chains* may be important in trying to explain – and eventually prevent – the occurrence of a disease.

Finally, I would like to draw your attention to synergy between causal factors. If *A* and *B* are causal factors for a disease, and the incidence increases from 0.01 to 0.02 in the presence of *A* and from 0.01 to 0.03 in the presence of *B*, then it would be expected that the presence of both factors (*A* and *B*) would increase the incidence from 0.01 to 0.04 if their effects were additive. This is sometimes the case. Quite often, however, the increase in incidence in the presence of two causal factors is greater than would be expected from additivity. *Synergy* may be defined as the relationship between factors which exhibit a joint effect that exceeds the sum of their separate effects. For example, both smoking and asbestos exposure increase the incidence of lung cancer, but smoking has a greater effect – in terms of increase in lung cancer incidence – among asbestos workers than among those unexposed to asbestos.

Keeping these empirical observations in mind, let us move on to a model (see Figure 1) that is well known to a generation of epidemiologists. It shows a causal factor, here called the *agent*, and seems to illustrate that the effect of this agent is influenced by the *host*, i.e., the person exposed to the agent, and by his *environment* [1]. This model was developed during a period when much attention was directed towards infectious disease epidemiology. The "agent", of course, represents some kind of bacteria or other microorganism. The importance of this model (except that it is familiar

ENVIRONMENT

Fig. 1. Agent-host-environment model.

to so many epidemiologists) may be that it points out that the effect of an infectious agent depends on characteristics of the individual and his environment. Examples of the former are nutritional status, specific immunity and certain anatomical and physiological characteristics. Examples of the latter are population density, climate and fauna (particularly the presence of so-called vectors).

Figure 2 shows another model, called the *web of causation* [2], which is here used to illustrate the causation of myocardial infarction. The web of causation is different from the previous model, in that several causal factors – in the "host" as well as his environment – are identified. Furthermore, there are some indications of the relationship between these causal factors.

Causal chains, like the one I mentioned previously, are a basic element in the web of causation. In these chains, some factors (at the top of the figure) are more indirectly related to the outcome (myocardial infarction) than others. The figure seems to indicate that some factors, like coronary occlusion, are directly related to the disease. However, just as coronary atherosclerosis may lead to myocardial infarction by causing coronary occlusion, coronary occlusion may lead to myocardial infarction by causing lack of oxygen supply to the heart muscle. Hence, the relationship between coronary occlusion and myocardial infarction would be indirect. In most situations, I believe, it would be possible to think of intermediate factors

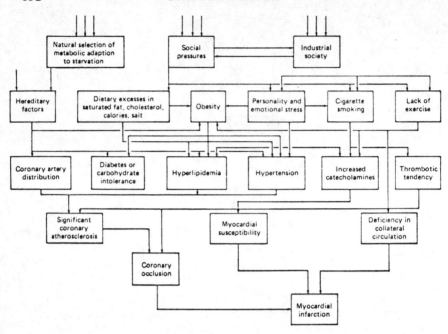

Fig. 2. Web of causation for myocardial infarction (from Friedman, 1974).

making the relationship between a causal factor and the disease indirect.
From an epidemiological point of view, and particularly when it comes to
applications of epidemiology in preventive medicine, the most interesting
factors may be found high up in the web of causation: in this case for example dietary factors, stress, smoking, and lack of exercise.

In the web of causation, different causal chains are connected by causal
relationships between factors on different levels. For instance, hypertension
may be due to hereditary factors, dietary habits and emotional stress. Lack
of exercise may cause obesity, but also deficiency in collateral circulation.
Causal chains, it seems, are not isolated but related to each other in a complex way. These interrelationships are, of course, important in understanding
the impact of a single causal factor or a certain combination of factors.

The *sufficient/component cause model* shown in Figure 3 illustrates the
causation of a hypothetical disease in the population [4]. This disease has
3 different sufficient causes, each consisting of 5 causal factors like pieces
of a pie. Some of these factors (like *E*) are only part of one sufficient cause,
while others (like *B*) appear in more than one sufficient cause. In this case,

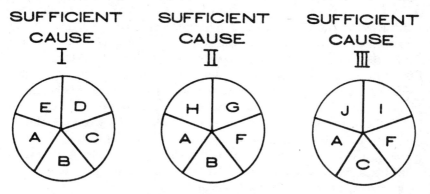

Fig. 3. Sufficient/component cause model (from Rothman, 1976).

one factor *A* appears as a member of all sufficient causes. Hence, the presence of *A* is a necessary condition for the disease.

As suggested by this model, identification of all the components of a given sufficient cause is unnecessary for disease prevention, since blocking the causal role of but one component of a sufficient cause renders the joint action of the other components insufficient and prevents the effect. For instance, the elimination of factor *E* (Figure 3) from the population would prevent all cases of the disease attributable to sufficient cause *I*.

In epidemiology, a causal factor is sometimes said to be "weak" if its presence confers only a small increment in disease occurrence, but "strong" if its presence increases the occurrence of disease substantially.In Figure 3, factor *E* will cause the disease only in the presence of factors *A-D*. Furthermore, the extent to which the presence of *E* causes the disease is determined by the occurrence of factors *A-D* in the population. For example, in a society where a certain kind of atmospheric pollutant is very common, the rare inheritance of a gene causing allergy against it would appear to be a "strong" causal factor for that disease, but exposure to the pollutant would appear to be a weak causal factor. In another society, however, in which this pollutant is rare, but inheritance of the gene is very common, inheritance would be a weak causal factor and exposure to the pollutant would be a strong causal factor. Hence, according to this model, the "strength" of a causal factor depends on the occurrence of the complementary component factors in the same sufficient cause.

Synergy between causal factors was previously described on the basis of empirical observations. In this model, synergy implies that two causal factors are members of the same sufficient cause. On the other hand, if two causal factors are components of different sufficient causes, and are not members of any other sufficient cause, they will show no synergy. Of course, two causal factors may well be components of different sufficient causes but also coexist as members of a third sufficient cause. Such "partially" synergistic effects are in agreement with what has been found in epidemiological studies of many diseases. One example is the effect of smoking and alcohol consumption on the occurrence of mouth and pharynx cancer. Either of these factors seem to increase the occurrence of such cancer, but their combined effect exceeds the sum of their individual effects. According to the model, necessary causes — which are members of all sufficient causes — are synergistic with all other causal factors.

When a causal factor is seen in relation to the occurrence of a disease in a particular population, its public health importance will depend upon the fraction of the disease occurrence which results from the sufficient cause (or causes) to which the causal factor belongs. This has been called the *etiologic fraction* of a causal factor. In Figure 3, the etiologic fraction of factor E is equal to the proportion of cases of the disease attributable to sufficient cause I, while the etiologic fraction of factor B is equal to the proportion of cases attributable to sufficient causes $I + II$. The sum of the etiologic fractions of all causal factors in Figure 3 will, of course, exceed 100%. Similarly, empirical data suggest that, for example, alcohol and tobacco as etiologic factors for mouth cancer each have an etiologic fraction greater than 50%. The etiologic fraction of a necessary causal factor will, of course, always be equal to 100%.

This is just a small sample of three out of many models of causation in epidemiology. Nevertheless, they may reflect — to some extent — how epidemiologists have been reasoning concerning the relationship between diseases and their causes. Epidemiologists are usually more concerned with empirical data than with theory (not to mention philosophy). Empirical observations, rather than theoretical considerations, are the basis from which these and other models of causation in epidemiology have grown. As more and more empirical data became available, and as the methods for collecting and analyzing such data were improved, this basis became more stable but also more complex. Furthermore, the field of investigation has been broadened considerably in recent decades, from mainly infectious disease epidemiology to cardiovascular diseases, cancers and other diseases. To some extent, the models shown illustrate this development.

In the first model one causal factor — the "agent" — is identified, while "host" and "environment" are thought of as modifying rather than causal factors. The second and third model, on the other hand, identify several causal factors related to the effect (disease) in a similar way. The relationship between causal factors is described in the last two models, although in different ways. The second model focuses on causal sequences, and the relationship between them. The third model describes causal factors as components of sufficient causes, illustrating such empirical observations as the "strength" of and synergy between causal factors, and their public health importance in terms of etiologic fractions.

These models, like others, are limited in that they focus on certain aspects of the relationship between diseases and their causes. This is their weakness, because other aspects may be lost or neglected. But this is also their strength, because it may help to focus our attention on important aspects of complex empirical data.

BIBLIOGRAPHY

[1] Evang, K.: 1974, *Helse og Samfunn*, Gyldendal Norsk Förlag, Oslo.
[2] Friedman, G. D.: 1974, *Primer of Epidemiology*, McGraw-Hill, London, U.K.
[3] MacMahon B. and Pugh, T. F.: 1970, *Epidemiology Principles and Methods*, Little Brown and Company, Boston, Mass.
[4] Rothman, K. J.: 1976, 'Causes', *American Journal of Epidemiology* 104, 587–592.

B. INGEMAR B. LINDAHL

ON THE SELECTION OF CAUSES OF DEATH:
AN ANALYSIS OF WHO's RULES FOR SELECTION
OF THE UNDERLYING CAUSE OF DEATH

1. INTRODUCTION

The mortality statistics, which are an important source for medical research, are prepared in many countries according to regulations and recommendations issued from the World Health Organization (WHO) in the current revision of the *Manual of the International Statistical Classification of Diseases, Injuries, and Causes of Death* (ICD). These statistics are generally based on *one* cause only per death, and the way to select this cause is determined by WHO through certain rules in the ICD.

WHO's criteria and rules for selection of this principal cause are not sufficiently known among physicians, although these statistics are utilized in epidemiological research as an instrument to get information about important risk-factors. They are also utilized for international comparisons and as an important factor in public health planning, i.e., what preventive measures should be adopted in the future.

The purpose of this essay is to shed light on these rules as formulated by WHO in the recent revision of the ICD [6] : to examine how they can be interpreted and what criteria of selection may be derived from them. The basis of analysis is mainly philosophical.

2. THE MEDICAL CERTIFICATION FORM AND THE
DEFINITION OF UNDERLYING CAUSE OF DEATH

According to WHO, the main purpose of the ICD rules is to bring forth causes of death in the form of data which can be used to decide what is to be prevented, and thereby most effectively prevent untimely death. This purpose is stated in the ICD, where the WHO defines the so-called *underlying cause of death*.

From the standpoint of *prevention of deaths*, it is important to cut the chain of events or institute the cure at some point. The most effective public health objective is to prevent the precipitating cause from operating. For this purpose, the underlying cause has been defined as "(a) the disease or injury which initiated the train of morbid events leading directly to death, or (b) the circumstances of the accident or violence which produced the fatal injury." (My italics [6], pp. 699–700.)

L. Nordenfelt and B.I.B. Lindahl (eds.), Health, Disease, and Causal Explanations in Medicine, 137–152.
© 1984 *by D. Reidel Publishing Company.*

The selection task is imposed upon the attending physician as he or she signs the medical certificate of death. It is incumbent on the physician to identify *the underlying cause of death* and to record the principal course of events from death back to the underlying cause of death, and also *any condition contributing to death* but not directly part of the fatal sequence. For this purpose WHO designed in 1948 a form of certification of all causes of death, the International Form of Medical Certificate of Cause of Death (Figure 1).

INTERNATIONAL FORM OF MEDICAL CERTIFICATE OF CAUSE OF DEATH

Cause of Death		Approximate interval between onset and death
I		
Disease or condition directly leading to death *	(a)................ due to (or as a consequence of)	....
Antecedent causes Morbid conditions, if any, giving rise to the above cause, stating the underlying condition last	(b).................. due to (or as a consequence of) (c).................	....
II		
Other significant conditions contributing to the death, but not related to the disease or condition causing it	....	....
* This does not mean the mode of dying, e.g., heartfailure, asthenia, etc. It means the disease, injury, or complication which caused death.		

Fig. 1. (Reproduced from [6].)

As it appears from the figure, the section of the certificate where the physician states his or her opinion about the conditions leading to death consists of two parts. In Part I the physician describes the main chain of events, with *one* cause on each line from (*c*) to (*a*). Other significant conditions are stated in Part II. The physician may omit the lines (*b*) and (*c*), "if the disease or condition directly leading to death stated in line (*a*) describes completely the train of events" ([6], p. 700).

The underlying cause of death selected by the physician will be selected as the underlying cause by the statistical office for primary tabulation, unless this assignment deviates from provisions of the ICD (e.g., entry of a highly improbable sequence). In this case, if further clarification of the certificate by the certifier has not been possible, the coder is entitled to *re-select* the underlying cause of death, from among the causes recorded on the certificate, in accordance with the rules in the ICD. Thus, in those instances where the certificate has not been properly completed these rules determine the selection of the underlying cause for primary mortality tabulation purposes.

3. WHO's RULES FOR SELECTION OF THE UNDERLYING CAUSE OF DEATH

When only *one* diagnosis is entered on the death certificate, and this diagnosis completely describes the train of events, according to the ICD this cause should be selected for tabulation. When *more* than one cause is recorded by the physician, selection of the underlying cause of death should be made in accordance with the ICD rules.

There are three kinds of ICD rules: *Selection rules, Modification rules* and *Notes for use in underlying cause mortality coding.*[1] The first two kinds of rules amount to 12; these are completely described below. The *Notes* amount to 62. To a great extent these *Notes* may be regarded as mere applications of the *Selection* and *Modification rules*. For the main part I will therefore confine myself to the 12 principal rules.

The selection of the underlying cause of death is made in two steps; first the cause is selected according to the *Selection rules*; then this cause is modified according to the *Modification rules*.

There are four *Selection rules*:

i. Selection Rules

General rule. Select the condition entered alone on the lowest used line of Part I unless it is highly improbable that this condition could have given rise to all the conditions entered above it.

Rule 1. If there is a reported sequence terminating in the condition first entered on the certificate, select the underlying cause of this sequence. If there is more than one such sequence, select the underlying cause of the first-mentioned sequence.

Rule 2. If there is no reported sequence terminating in the condition first entered on the certificate, select this first mentioned condition.

Rule 3. If the condition selected by the *General rule* or *Rules 1 or 2* can be considered a direct sequel of another reported condition, whether in Part I or Part II, select this primary condition. If there are two or more such primary conditions, select the first mentioned of these. ([6], pp. 701–702).

The *General rule* should be disregarded and the other *Selection rules* applied only if *more than one condition* is entered on the last used line in Part I or, when only one condition is entered on this line, it is *highly improbable* that this condition could have given rise to all the conditions entered above it, and (in both cases) only if *further information* from the certifier to clarify the certificate has not been available ([6], p. 702). Otherwise, the diagnosis entered on the last entered line in Part I should be selected as the underlying cause of death.

ii. Modification Rules

When the underlying cause of death has been selected according to the *Selection rules*, the *Modification rules* should be applied. These rules are intended to improve the usefulness and precision of mortality tabulations ([6], p. 703).

Some of the *Modification rules* require a renewed application of the *Selection rules*.

There are nine *Modification rules*:

Rule 4. Senility. Where the selected underlying cause is classifiable to 797 (Senility) and a condition classifiable elsewhere than to 780–799 is reported on the certificate, re-select the underlying cause as if the senility had not been reported, except to take account of the senility if it modifies the coding.

Rule 5. Ill-defined conditions. Where the selected underlying cause is classifiable to 780–796, 798–799 (the ill-defined conditions) and a condition classifiable elsewhere than to 780–799 is reported on the certificate, re-select the underlying cause as if the ill-defined condition had not been reported, except to take account of the ill-defined condition if it modifies the coding.

Rule 6. Trivial conditions. Where the selected underlying cause is a trivial condition unlikely to cause death, proceed as follows:

(a) if the death was the result of an adverse reaction to treatment of the trivial condition, select the adverse reaction.

(b) if the trivial condition is not reported as the cause of a more serious complication, and a more serious unrelated condition is reported on the certificate, re-select the underlying cause as if the trivial condition had not been reported.

Rule 7. Linkage. Where selected underlying cause is linked by a provision in the classification in the Notes for use in primary mortality coding on pages 713–721 with one or more of the other conditions on the certificate, code the combination.

Where the linkage provision is only for the combination of one condition specified as due to another, code the combination only when the correct causal relationship is stated or can be inferred from application of the selection rules.

Where a conflict in linkages occurs, link with the condition that would have been selected if the underlying cause initially selected had not been reported. Apply any further linkage that is applicable.

Rule 8. Specificity. Where the selected underlying cause describes a condition in general terms and a term which provides more precise information about the site or nature of this condition is reported on the certificate, prefer the more informative term. This rule will often apply when the general term can be regarded as an adjective qualifying the more precise term.

Rule 9. Early and late stages of disease. Where the selected underlying cause is an early stage of a disease and a more advanced stage of the same disease is reported on the certificate, code to the more advanced stage. This rule does not apply to a "chronic" form reported as due to an "acute" form unless the Classification gives special instructions to that effect.

Rule 10. Late effects. Where the selected underlying cause is an early form of a condition for which the Classification provides a separate late effects category and there is evidence that death occurred from residual effects of this condition rather than in its active phase, code to the appropriate late effects category.

The following late effects categories, including those in the Supplementary E code, have been provided: 137, 138, 139, 268.1, 326, 438, 905–909, E929, E959, E969, E977, E989, and E999. (See III Late effects, page 723).

Rule 11. Old pneumonia, influenza and maternal conditions. Where the selected underlying cause is pneumonia or influenza (480–487) and there is evidence that the date of onset was 1 year or more prior to death or a resultant chronic condition is reported, reselect the underlying cause as if the pneumonia or influenza had not been reported. Where the selected underlying cause is a maternal cause (630–678) and there is evidence that death occurred more than 42 days after termination of pregnancy or a resultant chronic condition is reported, reselect the underlying cause as if the maternal cause had not been reported. Take into account the pneumonia, influenza or maternal condition if it modifies the coding.

Rule 12. Errors and accidents in medical care. Where the selected underlying cause was subject to medical care and the reported sequence in Part I indicates explicitly that the death was the result of an error or accident occurring during medical care (conditions classifiable to categories E850–E858, E870–E876), regard the sequence of events leading to death as starting at the point at which the error or accident occurred. This does not apply to attempts at resuscitation. ([6], pp. 706–712).

iii. Notes for Use in Underlying Cause Mortality Coding

Of the 62 *Notes*, about half (27) can be regarded as mere applications of the *Selection rule 3* and the *Modification rules 7–9.*[2] The rest can be divided into three categories (*a–c*):

(*a*) Rules for selection between a more and a less serious condition. The rules give priority to more serious injuries and malignant conditions over slighter injuries and benign conditions. These rules resemble *Rule 6* as they attach importance to the seriousness of the cause.

(*b*) Rules for complications of pregnancy, childbirth and the puerperium. In cases of complications due to fetopelvic disproportion one *Note* gives priority to a pelvic abnormality diagnosis. "Fetopelvic disproportion" is according to the ICD a "disproportion of mixed maternal and fetal origin, with normally formed fetus" ([6], p. 365). I take this to mean that the "fetopelvic disproportion" diagnosis refers to cases when the incongruence in size and/or shape between fetus and birth canal is due to *not only* pelvic obstruction *but also* to fetal size and/or malposition. Whereas "pelvic abnormality" refers *only* to some abnormality of pelvis as the cause of the complication. Thus, a less informative diagnosis, "pelvic disproportion", is given priority over a more informative diagnosis, "fetopelvic disproportion". This constitute a deviation from the general purpose of the re-selection (viz, to attain "the most useful and *informative* condition for tabulations of mortality data".) (My italics [6], p. 703.)

In conformity with this *Note* another *Note* gives priority to "obstruction by bony pelvis" over the fetus-oriented diagnosis, "obstruction caused by malposition of fetus at onset of labour", if both are mentioned on the certificate.

A third *Note* gives precedence to conditions originating in the perinatal period (i.e., all conditions in category XV of the ICD) over residual cerebral paralysis at ages 4 weeks or over; e.g., symptomatic torsion dystonia (ICD code 333.7) and infantile cerebral palsy (ICD code 343).

(*c*) Further rules. This rest-category contains rules to the effect that certain psychoses (ICD codes 293–294, 299.1), or specific nonpsychotic mental disorders following organic brain damage (ICD code 310) or mental retardation, should not be used as the underlying cause of death if the underlying *physical* condition is known.

Epilepsy related to an injury is treated in a speical *Note*: When an accident results from an epileptic seizure, epilepsy should be selected; when the epilepsy is due to trauma the nature and/or the cause of the injury should be selected.

Diagnoses of postsurgical complications, such as postsurgical hypothyroidism or hypoinsulinaemia, functional disturbances following cardiac surgery, postmastectomy lymphoedema syndrome, or others (ICD codes 244.0, 251.3, 383.3, 429.4, 457.0, 564.2, 564.3, 569.6, 576.0, 579.2, 579.3) should not be selected if the reasons for the surgery is known. In these cases, the condition for which the surgery was undertaken should be given priority over the postsurgical complication. If these *Notes* are not meant to be exceptions to *Rule 12*, they presuppose that the postsurgical complication was not due to any error or accident in the surgery.

In cases of bronchitis, emphysema or ischemic heart disease due to "nondependent abuse of tobacco", any of the first three diagnoses should be given priority over the abuse of tobacco. This rule constitute an exception from the *Selection rule 3*, in that priority is given to a condition closer to death.

4. TWO DIMENSIONS OF CAUSAL MULTIPLICITY

Death is preceded by causes mainly in two ways. Two causes A and B can be (one way) *interdependent* parts of a succession of causes leading to death (dimension 1), or they can be *non-interdependent* contributary causes (dimension 2).[3] In the first dimension the two causes are separated in time (e.g., when a car accident causes a serious fracture of skull), in the second dimension they can be simultaneous (e.g., when two deadly injuries are inflicted at the same time). Although attention is not explicitly paid to this distinction in the ICD, separate criteria are provided for selection in the two dimensions, and it is also reflected in a conceptual discrepancy between WHO's definition of the underlying cause of death and how this concept evolves from the ICD rules.

4.1. Dimension One

On the medical certificate form the physician is requested to describe the (main) chain of events leading to death in three steps: "(*a*) *Direct cause* (due to) (*b*) *Intervening antecedent cause* (due to) (*c*) *Underlying antecedent cause*," (My italics [6], p. 700.)

According to this basis of division the physician should select the under-lying cause of death with regard to its *position* in the chain of events; i.e., the point where the sequence leading to death started.

This initiating (underlying antecedent) cause (*c*) is of course in turn always preceded by a sequence of events (*x, y, z, ... , n*), but as these can't be re-garded as parts of "the sequence directly leading to death", none of them can be selected as the underlying cause of death.

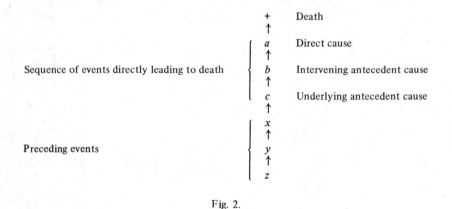

Fig. 2.

WHO's definition of the underlying cause of death takes into account only this first dimension: the underlying cause is defined as the first pathological point or the external cause (circumstances) of the fatal sequence.

The definition concerns only *one* causal sequence and states which single, sequential cause should be chosen as the underlying cause of death.

Certain of the rules specify and elucidate the WHO definition of the underlying cause of death, for example *Rule 6*, which states that the under-lying cause of death must not be a condition which is unlikely to cause death or which is not reported as the cause of a more serious complication.

According to *Rule 9*, if more than one stage of the same disease is men-tioned, the most advanced stage should be selected as the underlying cause. These two rules limit the concept "the underlying cause of death" to the first *severe* or *advanced* condition in the chain of events leading to death. Thus the very first pathological condition in the sequence is not always regarded as the initiating condition.

Other rules, as for example *Rules 1, 3* and *6*, exceed the definition of the underlying cause of death and guide the choice between competing initiating causes and alternative causal chains directly leading to death (dimension 2).

4.2. Dimension Two

Rules 1, 3 and *6* exceed the definition of underlying cause of death as they direct selection between causes in more than one sequence (*Rule 1*), between two or more primary conditions (*Rule 3*), as well as between unrelated conditions (*Rule 6*).

Thus, the ICD rules give guidance in both dimensions, partly to make it possible to distinguish fatal sequences from other preceding events (dimension 1; Figure 2), and partly to enable the selection of *one* initiating condition among several simultaneous or otherwise independent initiating causes (dimension 2; Figure 3).

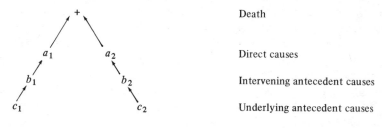

	Death
a_1 a_2	Direct causes
b_1 b_2	Intervening antecedent causes
c_1 c_2	Underlying antecedent causes

Fig. 3.

Selections in dimension 2 (for instance between c_1 and c_2 in Figure 3) can be made either *arbitrarily*, as in *Rule 1* and *3* where the first-mentioned cause is selected (e.g., c_1), or it can be made according to the *weight* of the cause, as in *Rule 6* where the most serious cause is selected.

Thus the initial *position* in each causal chain qualifies a cause to be regarded as *an* underlying cause of death (i.e., an *initiating* cause of death); the *weight* decides which of these initiating causes should ultimately be selected as *the* underlying cause of death (i.e., the *most important* initiating cause of death).

5. VARIOUS FUNCTIONS OF THE RULES

The rules of WHO have various functions. Some are *semantic* rules which concern the diagnosis proper, not the cause itself. Some are *causal* rules, which allocate priority with regard to characteristics of the cause. Other rules are *purely conventional*; they neither refer to the semantic character of the diagnosis nor to the characteristics of the cause itself, but merely present purely conventional directives to facilitate the work of the coder.[4]

i. Semantic Rules

The *Modification rules 5, 7* and *8* are semantic rules, which give priority to more precise (*Rule 5*), inclusive (*Rule 7*), and specific (*Rule 8*) expressions. *Notes* which can be regarded as simply applications of the *Modification rules 7* and *8*[5] are also semantic rules.

ii. Causal Rules

The main part of the ICD rules are in some sense causal rules. The *General rule* calls for a causal judgment of whether it is probable that the cause in question could have given rise to all the conditions entered above it. *Rule 1* is only partly a causal rule. Although it gives priority to "the underlying cause" which is a causal concept, it is also conventional, since it gives priority to the "first-mentioned" sequence. *Rule 3* and those *Notes* which can be regarded as direct application of *Rules 3*,[6] give priority to the more primary condition. Thus they can be regarded as causal rules.

   Several rules give priority to explicitly more serious conditions over less serious ones, which is a causal characteristic. (*Rule 6*, and the two *Notes* cited in the beginning of Chapter 6 below).

   In the same manner *Rules 9–11* are concerned with a causal characteristic — the stage of the disease (i.e., early or late) — and if death occurred from later effects of the disease.

   Rule 9 can be regarded as an example of a rule which is concerned with the seriousness of the cause, whereas *Rules 10–12* are concerned with the position of the cause — where in the main sequence the cause is situated (*Rule 10* and *11*) — and whether the cause is a result of an error or accident occurring during medical care. Thus, all four rules can be regarded as causal rules.

iii. Purely Conventional Rules

Selection rules 1–3 are purely conventional rules. According to *Rule 1*, when selecting between underlying causes in two separate chains of events (e.g., c_1 and c_2 in Figure 3) priority is given to the *first-mentioned* one (i.e., c_1).

In a similar way *Rule 2* gives priority to the *first-entered* condition, when there is no reported sequence terminating in this condition. When selecting between two or more primary conditions (whether in Part I or II on the death certificate), *Rule 3* gives priority to the *first-mentioned* condition.

6. THREE PRINCIPAL CRITERIA

As mentioned, *Rule 6* and two *Notes* give explicit priority to a *more serious* condition. The *Notes* state:

035 *Erysipelas*, 037 *Tetanus*, 038 *Septicaemia*.
Code to these diseases when they follow vaccination or a slight injury (any condition in N910–N919, prick, splinter, minor cut, puncture (except of trunk), bruise or contusion of superficial tissues or external parts, burn of first degree); when they follow a *more serious* injury, code to the injury. (My italics [6], p. 713.)

460 *Acute nasopharyngitis* (*common cold*), 465 *Acute upper respiratory infections of multiple and unspecified sites*.
Excludes when reported as the underlying cause of *serious* conditions such as meningitis (322.9), intracranial abscess (324.0), otitis media (381, 382), mastoiditis and related conditions (383), pneumonia and influenza (480–483, 485–487), bronchitis and bronchiolitis (466, 490, 491), acute nephritis (580.0–580.9). (My italics [6], p. 718.)

These and other *Notes* as well as *Rule 9* can be regarded as selecting the underlying cause of death according to a criterion, which I choose to call *the criterion of severity* (e.g., "a more advanced stage" in *Rule 9* can be interpreted to mean a more serious stage).

This criterion is a central criterion in the ICD; it has been a recurrent criterion in manuals even before the ICD, since the first known list of rules for selection of the primary cause of death which dates from the end of the 19th century [4].

Another criterion, which derives from the purpose of the ICD rules, is what we can call *the criterion of prevention*.

According to Collingwood [1] this criterion is common to selection of "the cause" in natural science (e.g., in medical science): "The question 'what is the cause of an event y?' means in this case 'how can we produce or prevent y at will?' ". Collingwood's criterion includes both prevention and production of the cause in question. Let us call this criterion *the criterion of manipulability*. According to Collingwood, "the cause" of an event in natural science is *by definition* the (most) manipulable cause. Thus, according to Collingwood, this criterion applies without exception in natural science. He states:

Suppose someone claimed to have discovered the cause of cancer, but added that his discovery though genuine would not in practice be of any use because the cause he had discovered was not a thing that could be produced or prevented at will. Such a person would be universally ridiculed and despised. No one would admit that he had done what he claimed to do. It would be pointed out that he did not know what the word cause (in the context of medicine, be it understood) meant. For in such a context a proposition of the form 'x causes y' implies the proposition 'x is something that can be produced or prevented at will' as part of the definition of 'cause' ([1], p. 90).

As we have seen, Collingwood's observation is germane to the WHO's recommendations for the selection of the underlying cause of death in the ICD. Although manipulability neither is a criterion of *definition* of the underlying cause of death in the ICD, nor is there anything that speaks in favour of a necessity to *produce* the cause selected, it still seems as if prevention is a central *empirical* criterion of selection in the ICD.

According to the ICD the underlying cause of death should be such that the prevention of this cause is the most effective way to prevent untimely death. Thus, the merit of the preventability of the underlying cause is dependent on the efficiency by which the prevention of this cause prevents death. In a situation where two causes *A* and *B* are equally easy to prevent, but where the prevention of *A* more effectively prevents death, *A* is regarded as a more important cause than *B*. Whereas if the prevention of *A* prevents death as effectively as the prevention of *B*, but *A* is more easily prevented, then *A* is considered a more important cause than *B*. Consequently, according to this criterion, the most important cause to be selected would be the cause which is *both* the easiest to prevent and the prevention of which would be comparatively the most effective way to prevent death.

A crucial question here is, of course, to what extent these two criteria (the criteria of severity and prevention) are compatible. Both "severity" and "prevention" are vague concepts with manifold meanings, and therefore in need of further penetration.[7] (Although I shall not be able to go into this analysis here, I would like to mention a rather common use of "severity" in medicine, where a severe condition simply could mean *an irreversible* (*lethal*) *condition*.) If we mean by prevention what Morris [3] calls *secondary prevention* (i.e., prevention of progress of disease [3], p. 134), these two concepts are very obviously incompatible. The priority of conditions of *a more advanced stage* in *Rule 9* can be interpreted in a way that actualizes this potential conflict.

We saw that according to Collingwood the criterion of manipulability is both a necessary and sufficient criterion for selection of "the cause" in natural science. However, Hart and Honoré [2] argue against this position: "The discovery of the cause of cancer would still be the discovery of the cause, even if it were useless for the cure or avoidance of the disease ..." ([2], p. 34). According to Hart and Honoré, the criterion of manipulability is subordinate to another more general criterion which we can call *the criterion of abnormality*: what is to be counted as the cause of an event (e.g., a person's death) should be something which *makes the difference* from what usually characterizes the phenomenon in question, either as a particular object (e.g., a particular person) or as a type of object (e.g., human beings in general). According to Hart and Honoré a selection of the cause which is not accomplished in accordance with this criterion lacks explanatory value.

When we select the manipulable cause it is not solely because of the manipulability of the cause, according to Hart and Honoré, but because experience has forced us to accept the non-manipulable cause as the natural course of events: "... for the factors which we cannot control may persist and, however unwelcome, become known and accepted as the normal course of things, as the standing conditions or environment of our lives to which we adjust ourselves more or less well" ([2], p. 34). The criterion of abnormality is in these cases a principal criterion, according to Hart and Honoré.

WHO approaches the criterion of abnormality (in a special sense of this criterion) when it is stated in the ICD that the concept of the underlying cause of death does not include *modes of dying* ([6], p. 699). The "modes of dying" can be interpreted as the processes which *always* precede death or through which death *always* ensues. According to Hart and Honoré, pointing out such a cause would be of no interest in applied sciences, as for example in clinical medicine, which seeks to explain *particular* events (e.g., why a particular person P died just at the point of time t under the circumstances c) and not how death *generally* ensues. They give the following example:

Take, for example, a case where one man shoots another and kills him. Here we should treat the shooting, not the later deprivation of his blood-cells of oxygen as the explanation and the cause of his death, although it is perfectly true that we could predict the man's death from knowledge of the latter more certainly than we could from knowledge of the earlier part of the process ([2], pp. 36–37).

The WHO's qualifier about "modes of dying" can be understood as an attempt to avoid an explanation which would be *useless* for understanding the particular cause of death in the individual case; this tends to underscore the importance of the criterion of abnormality for the selection of the underlying cause of death.

In the final analysis what we refer to as "modes of dying" or "causes of death" depends on how we define "death". Apparently we have to separate this concept sharply from our definitions of modes of dying and causes of death.

7. SUMMARY

WHO's definition of the underlying cause of death only partly reflects the connotation of this concept as it evolves in the ICD. The definition restricts the concept of the underlying cause of death only in relation to other causes in the same chain of events, whereas the WHO's rules for selection of the underlying cause of death also identify the underlying cause of death among mutually independent causes, in, for instance, different parallel chains of events.

Thus we have been able to discern two dimensions of multiplicity of causes of death, i.e., (one way) *interdependent* and *non-interdependent* causes, the WHO definition of the underlying cause has only considered the first dimension. Since the rules for selection of the underlying cause of death are not included in the instructions for physicians issued by the WHO [7], it is not possible for the physician to understand fully the WHO concept of the underlying cause of death.

We have discerned two criteria of selection between causes in both these dimensions, viz, what we have called *the criterion of severity* and *the criterion of prevention*. We have discussed their interpretation and applicability, and we have found that in one interpretation of these criteria, they are not compatible, viz, when severity is understood as an irreversible (lethal) condition and preventions as prevention of progress of disease.

Their explanatory value has been discussed with reference to the position of Hart and Honoré's [2] about the general application of a third criterion in medicine, viz, *the criterion of abnormality*. We have found that the ICD assumes an additional application of this criterion when selecting the underlying cause of death.

Huddinge University Hospital,
Sweden

ACKNOWLEDGEMENTS

I am indebted to Dr Lennart Nordenfelt for valuable criticism and inspiration of ideas in the development of this essay. I have particularily benefited from his essay *Causes of Death* [4].

NOTES

[1] Two additional kinds of rules should be mentioned here: (a) *Rules embedded* in the text in other parts of the ICD; e.g. the rule that says that the physician may omit the line (*b*) and (*c*) and enter only one cause on the certificate, if this single diagnosis "describes completely" the train of events leading to death ([6], p. 700): and (b) *Meta-rules*, which state when and how the *Selection-, Modification rules*, and the *Notes* should be applied.
   However, for the sake of simplicity I shall omit these rules in this inquiry.

[2] The *Notes for use in underlying cause mortality coding* are not numbered in the ICD; but if we investigate them in the order they are mentioned in the ICD ([6], p. 713–721), we can make the following comparisons between the *Selection-* and *Modification rules* (S&M) and the *Notes* according to a hypothetical numbering of the *Notes*:

S&M		Notes
3	=	10, 15, 16, 18, 29, 50, 60–62
7	=	33, 42, 43, 46, 49
8	=	1, 3, 7, 8, 20–22, 26, 32, 35, 41
9	=	40, 48

[3] This distinction has previously been described by White [5] and Nordenfelt [4] as "horizontal" and "vertical" causal multiplicity.

[4] All rules are of course conventional. But the rules I refer to as "purely conventional" are not supported by any reasons what so ever. They are only meant to facilitate the work of the coder.

[5] See Note 2.

[6] See Note 2.

[7] For a further analysis of the concept of severity see Nordenfelt [4]; for the concept of prevention see Morris [3].

BIBLIOGRAPHY

[1] Collingwood, R. G.: 1938, 'On the So-Called Idea of Causation', *Proceedings of the Aristotelian Society* 38, pp. 85ff.
[2] Hart, H. L. A. and Honoré, A. M.: 1959, *Causation in the Law*, Oxford, U.K.

[3]   Morris, J. N.: 1964, *Uses of Epidemiology*, Edinburgh, London U.K.
[4]   Nordenfelt, L.: 1983, *Causes of Death, A Philosophical Essay*, Report 83:2, Swedish Council for Planning and Coordination of Research, Stockholm, Sweden.
[5]   White, M.: 1965, *Foundations of Historical Knowledge*, New York, London.
[6]   WHO: 1977, *Manual of the International Statistical Classification of Diseases, Injuries, and Causes of Death*, Vol. 1, 9th revision, Geneva, Switzerland.
[7]   WHO: 1979, *Medical Certification of Cause of Death, Instructions for Physicians on Use of International Form of Medical Certificate of Cause of Death*, Geneva, Switzerland.

ØIVIND LARSEN

DISEASE FROM A HISTORICAL AND SOCIAL POINT OF VIEW: SOME REMARKS BASED ON THE NEEDS OF PREVENTIVE MEDICINE

1. INTRODUCTION – CAUSALITY AND PREVENTION

The theoretical interest in causality in medicine has practical dimensions and close ties to day-to-day work in several parts of medicine. These questions are extremely important to those members of the medical profession who are occupied within different preventive disciplines. Here, knowledge of causality chains has to be transformed into chains of counteraction. In this process definitions of health and disease, assessment of the historical and social implications of diseases and health risks, and the relationship between the diagnostic entities, their contents and consequences, play an important role.

Such definitions also have an important function in the relationship between medicine and society. They clarify which problems are to be regarded as medical and which are not. This means that they tell us when a problem is to be tackled under the umbrella of medicine – as a medical task and with medical tools – and when it is not. Definitions, terms and concepts indicate when medical responsibility ends, even when the chain of causality obviously extends far beyond this point.

2. THE PROBLEM OF CAUSALITY IN PUBLIC HEALTH

Let us for a moment transfer ourselves from the philosophical discussion to a position where the results of the debate may have immediate implications in terms of finances, of restrictions on human activity or simply for personal, medical advice which has to be given. The medical officer in charge of a State Department of Health has the tailoring of a medical policy as one of his paramount tasks. However, his colleagues serving a county, a district, in the municipal health services or who are responsible for the medical counselling in an industrial company also have the quest for causality right on their desks.

Preventive work is founded on causality and the pursuing of causality chains. There is but little room left for a theoretical discussion when a certain microorganism is assigned to the corresponding infection. However, this surely is not the case when a nervous disease or an imminent hypertension may be traced back to working conditions or living habits.

153

L. *Nordenfelt and B.I.B. Lindahl (eds.), Health, Disease, and Causal Explanations in Medicine*, 153–164.
© 1984 *by D. Reidel Publishing Company.*

In the handling of causality problems in public health administration, difficulties arise in at least three fields.

One of the main problems is the way the health condition of the population presents itself and makes up the background against which the medical preventive work has to be mounted. Morbidity assumes different faces, dependent on time and place. An additional complication: The chameleon-like morbidity is not a defined entity that can be unambiguously described — the body to which the different faces belong, indeed has a ragged fur!

Further, general interest may fluctuate between degrees of severity of the same ailment, from the faintest symptoms to mortality. This reflects itself in the naming of the problem. An important part of the description is the medical diagnosis, the identification and isolation of a medical entity. As each diagnosis has consequences and prerequisites of its own, the contents and the use of diagnoses on different levels — both in the development of a disease and in the information available to the user of the diagnostic term as well — may have considerable importance for the efforts in striving towards what is regarded as more favourable public health. In the course of time, this use of terms may also undergo a qualitative sort of aberration: In the cultural setting given, new conditions may be introduced on the medical scene, others on the other hand are fading out of the medical range of definitions. Deviance may also be noted along the quantitative axis: slight symptoms taken for granted as general parts of life may enter the stage and present themselves as medical phenomena.

The third problem is the place of the disease in the causality chain. This chain stretches out from distant underlying causes to late and indirect consequences. Somewhere on this line lies the medical range, i.e., the disease and its proper handling with traditional methods of prevention and cure. This operational range, however, is not necessarily identical with that very section of the causality chain where intervention would be most likely to exert an optimal effect.

3. THE FACES OF MORBIDITY

Morbidity might be looked upon as a cloud outside your airplane window. Depending on the position of the aircraft and the direction of the beams of sunlight, your impression of it shifts. But as the plane plunges into it, you observe another quality — its borders may hardly be exactly described, even in terms of the first droplets or of the degree of transparency.

A cybernetic or ecological approach to the interplay between health and disease probably would have the (i) "true morbidity" as its natural objective [1] — a comprehensive notion clearly distinguishing the cloud from the surrounding skies. As a frame of reference to the medical historian, to the social historian, or in our case to the medical officer of preventive medicine, insight into the natural course and spread of a disease would render the ideal information. This true morbidity, however, may scarcely or never be observed. To the historian, morbidity is disguised in narrative sources as (ii) "perceived morbidity": morbidity as seen through the eyes of time and place. Back to the flight again: The cloud is observed just in this moment by you — and this is not the same as the simultaneous observation performed by the meterologically trained pilot. Correspondingly, the medical profession of today meets with the perceived morbidity through the complaints presented by the patients and the evaluation of the health problems in public opinion. To the public health officer, this public perception is of crucial importance to the acceptance of preventive precautions proposed, although his professional perception may be different.

The culturally coloured popular experiences of the evils and hazards of illness on the other hand do not refer to the true morbidity, but rather to an (iii) "influenced morbidity", as the notion of a certain illness, the pains, fears, and prospects of being sick, varies with the medical treatment and social handling of the particular condition at the time given. To stay with the allegory for a moment, the cloud outside the bull's eye looks like a product of nature, but it perhaps is not, because of the convections from the man-made industrial plant underneath. Here, tuberculosis may serve as the classical example of human transformation of morbidity and the subsequent attitudes towards it. Nowadays, an occasional new case of a phthisis, a renal or a skeletal manifestation of a tuberculous infection, is usually quite readily treated with the proper drugs. Even the most invariably fatal meningitis of former times has lost much of its menacing reputation. Nevertheless, the dangers of tuberculosis are still there, despite the artifical — and reversible — situation of successful cure as the rule and not the exception. To the health care system, the stronghold against tuberculosis is prevention; however, when the hazards of the disease have more or less disappeared from the public mind, the awareness of the causal factors and the general acceptance of the expense, restrictions and alertness of the control systems are jeopardized.

In medical records and health statistics, morbidity appears as (iv) "classified morbidity", subjected to the prevailing professional principles for nosology, i.e., for the describing of disease. To the medical historian, the basic principles for change within diagnostic classification in the course of time will be obvious. As an example, medical statistics of the 18th century are only seldom suitable for modern comparisons, since diagnostics of that time were to a larger extent based on symptoms and clinical manifestations, rather than on etiology [3]. So, the notification of a "fever" or a "diarrhoea" might be adequate and sufficient to the doctor of the time of enlightenment. To us, a diagnosis fixed at this stage of events or stage of explication is of limited use.[2]

A time span of two hundred years might seem to be of little interest to practical medicine. Though the diagnostic disharmony presents a major obstacle to the understanding of the role played by the diseases in the starting decades of the demographic transition of the Western countries, the trouble arising is by no means resticted to history; it also blocks any fruitful comparisons between public health in ancient Europe with the corresponding events and the similar situations in developing countries of today.

Less appalling, but of similar importance, are the diagnostic differences between various countries or between one decade and another. Such a detail as the composition of the reference population for the age-adjusting of morbidity rates may influence the picture emerging from international statistics of the size and severity of a health problem in time and place.

On the basis of "classified morbidity", the professional background, society runs its health services and weights medical needs against others. In this way the morbidity of the population is currently adjusted and influenced, so that the picture of the public health in fact rather unveils an (v) "administered morbidity", where the degree of social modelling very often is a direct function of the degree of social development in general.

4. DISEASE IN A SEQUENCE OF EVENTS

In preventive medicine, a simple diagram is very often used to clarify the principles of practical work. In this model, the development of a case of sickness is described as a function of time from the zero point when the individual is perfectly healthy but under the influence of harmful, disease-producing circumstances. During an interval of time, which may vary from an instant to decades, disease develops and gradually passes through the subclinical stage and gets into the traditional sphere of curative medicine. The line drawn in the diagram at this point may take different directions; the disease may disappear and the curve bend back to zero, it may show a fulminant course and the patient die, or the outcome may be something in between, healing with some sort of disablement.

As the approaches and techniques which have to be applied by preventive medicine change, depending on at which point on this line they are applied, a period suitable for "primary prevention" is indicated in the pre-morbid state, where causes of disease have to be identified, localized and defeated. In the subclinical period the "secondary prevention" − e.g., the screening for unidentified disease − finds its work, whilst the "tertiary prevention" in the phase where illness is overt, consists in the preventing of unwanted complications and consequences as far as possible.

In this model, the course of a disease presents itself through a series of events. Following an exposure of some kind, morbid changes take place in the individual and at some point this is detected; a diagnosis is attached to the process and the individual becomes a patient. The logic of the model is not cause but chronology.

Historically and socially the interests and skills in detecting a disease will show variations along the time axis of the model. As the diagnosis usually is the key to medical intervention, there will inevitably be considerable variations in the contents and practical effects of the diagnoses, depending on at what level of medical information and at what level of morbid development they are used.

4.1. The Diagnostic Basis in a Series of Events

On its way from cause to consequence a disease may be described in many different ways. Symptoms produced are obvious for classification not only in the eighteenth century but even today, especially when etiology is obscure − some neurological diseases and syndromes might be used as examples here.

When a tissue or an organ is affected, a diagnosis may be fixed to the morphological changes produced. The workers in the field of preventive medicine, however, would rather ask for the etiology. So would most people within modern curative medicine too, because the etiological diagnoses give better guidance for successful treatment than do morphological descriptions, which could be correctly assigned to different underlying causes.

Diagnostics currently used is based as a rule on a mixture of these principles and thus is more readily connected with the sequence of events in the case of being sick -- or even in the progress of scientific understanding of the disease − than with causality. Historically, a steadily proceeding process of revision and alterations of scientific and popular diagnostics is observed, probably in favour of the causality principle in the long-run.

4.2. Use and Function of Diagnostic Terms

Usually, medical terms and diagnoses are thought of as linked to medical personnel, indicating a certain homogeneity in application and contents. However, the concepts of health governing those habits of life which in return exert their influence of health, and which expose themselves as primary objectives to preventive medicine have as a rule a heavy non-medical overload. Prevalence studies of the perception of disease in a population, the frequency and nature of complaints, reveal that a feeling of uneasiness for reasons that are judged by the individual as medical, usually occurs in a major proportion of the groups examined. These complaints, vaguely described in terms of symptoms, morphology or perhaps badly understood causality, make up the health background for the individual, governing his preferences in many ways. However, such concepts of disease also attain administrative value, e.g. when systems of self-reported sick-leaves are introduced. Then the light complaints, the professionally unverified diagnoses acquire a new dimension: they enter statistics and they entitle the patients to social benefits in a new way.[3]

Quite apart from its administrative implications, self-diagnosis becomes a heavier social weighting as a consequence of the generally rising levels in the standards of living. This tendency is supported by the culturally apparent shift in the evaluating of well-being and absence of disease, upgrading evils formerly more or less found acceptable or taken for granted.[4]

Finally, non-medical use of, interpretations of or launching of ideas about medical terms are by no means restricted to the private sphere. When medical considerations are taken into account in political decisions, the non-medical influence should not be underestimated.

5. DISEASE IN A CAUSALITY CHAIN

From the viewpoint of causality it is obvious that the model visualizing the types of prevention is defective in its logic. The labeling of links in the causality chain, identifying the proper background for preventive intervention, is related to a certain stage not only in the process from being healthy to being sick or dead, but on another scale as well, which permits the logical organization and use of etiological knowledge for preventive purposes in a more comprehensive way. Although modern diagnostics is made suitable to clinical medicine, supplementary discussion is needed for medical prevention. Here, clinically-based diagnostics often is defective and insufficient.

A clinical diagnosis may have a varying relationship to the causality chain.

5.1. *Disease as a Result*

A case of cancer of the lung may serve as a proper example of the place of the traditional diagnosis in the causality chain. We observe, say, a boy of seventeen regularly smoking twenty-five cigarettes a day. From epidemiological evidence we know that this individual may — or may not — acquire a lung cancer in the course of the decades ahead. More correctly: we know that he is considerably more prone to suffer from this disease at some time in the future than his mate who is a non-smoker. The medical diagnosis, however, is linked to the last stage of the process; when pathology is evident and disease is present, medicine steps in with its traditions in therapy and authority.

In this example, the diagnosis is a tool for the correct treatment of the patient. A diagnosis of lung cancer usually is morphologically and etiologically specified as well; the identification of an adenocarcinoma of the bronchial tissue in a long-time heavy smoker adds further weight to the epidemiological evidence of the causality linking smoking and cancer of the lung.

The discrepancy between the diagnosis as a tool for curative medicine and as a tool for preventive intervention is appalling in this case. Identification of the problem "heavy smoker" as a medical problem, legitimating actions by means of medicine and its inherent social role, would be far more effective to prospective patients and to health economy as well, than the last stage clinical treatment.

An example from occupational medicine might further emphasize this point. In Norway, the diagnosis of an occupational disease was attached hitherto to some sort of a pre-defined end point of an etiological process. When the doctor made his diagnosis of, say, asbestosis on the X-ray photo or concluded that the patient's rashes were due to industrial exposure, the set of medical and social events linked to an occupational disease was put into motion. However, new legislation and a shift in general attitudes have altered this situation; a case of disease, suspected by a general practitioner or even by the patient himself, has to be regarded as due to the occupation until clues indicating otherwise are at hand.

This transfer of the main interest usually moves the diagnostics to lighter attacks — another and more unfamiliar level in the pathological process. This very often causes problems in medicine, simply because methods of traditional diagnostics no longer show the same precision in the new situation. New tests have to be elaborated to cope with the new applications.

In this case there is another shift — not only in the stages of events in the development of a disease, but in the causality chain as well — from the manifestation of an occupational disease to the occupational exposures producing them; a shift within both the ranges of prevention and the ranges of curative medicine.

5.2. Disease as a Cause

If we accept that the assessment of medical diagnoses takes place somewhere not only in the sequence of events when a disease develops, but also in the very long chain of causes following each other and extending into the faraway shades on both sides of what we normally regard as a medical problem, we also have to look at the case when a disease starts the process we want to prevent – a process not necessarily being organically linked to the disease or the diseased person at all.

Sometimes social problems and social difficulties have a disease as a background. When within the traditional sphere of medicine, the patient as a rule will be taken care of by social medicine when social consequences appear. Nevertheless, the problem may end up in social misery – an example again:

A patient of mine and his family, back in the 1960s, were running a small grocer's shop in the countryside. Because of a suspicion of tuberculosis, routine medical examinations had to be carried through and proper medical precautions undertaken. In this case, however, the thing really suffering was most likely the grocer's shop, the base of subsistence for the family, due to old prejudices against tuberculosis.

Here, and in other cases of disease where there are cultural overtones, a real social problem outside the medical sphere may be caused by a disease. In our example, the attitudes of the surrounding population ought to have been subjected to "medical treatment" to avoid future difficulties for the patient. In this case too, the link of the causality chain which can be identified as the most suitable objective for intervention, lies beyond the range of diagnostics – and of medicine as currently perceived.

5.3. Cause Without Causality

A problem associated with modern diagnostics is the tendency to search for a cause where perhaps a causality chain is of minor practical importance. A diagnosis should also be able to describe an event in a sequence without causality links.

Based on international agreements, registration of deaths occurs according to forms where immediate and underlying causes of death are urgently asked for. Here, a fact like the inevitable cessation of a human life span or the biological variation attached to this fact, is paid only slight attention. Far back in the 18th century, when the first national bureau of statistics in the world commenced its notifications of deaths in Sweden in 1749, allowance was made for age and infirmity among the causes of death — truly valid diagnostic terms when mortality statistics had to serve the function of depicting an outline of the public health of the country, fit for health planning on the individual, regional, and nationwide scales.

6. CAUSALITY AND THE PRACTICAL FUNCTION OF THE CONCEPTS OF DISEASE

Looking back into the history of diseases and the history of medicine, it is obvious that our vision gets blurred by the fact that the concepts of what a disease is like, not only vary in the eyes of the contemporary spectators, but also in view of the fact that the concepts have a fluctuating relationship to the causality chain producing them. Within practical medicine, especially the preventive disciplines, this phenomenon causes concern and difficulties as well.

The "classified morbidity" presented to us is neither a homogeneous one over any span of time, nor does it tell us about the real interplay between diseases and individuals or populations. Due to the development of society, morbidity more and more becomes an "administrated morbidity", whilst the effect of this "administration" in the course of time will be measured through the information on "perceived morbidity", rather than e.g. in the traditional terms and figures of morbidity and mortality registration.

In a theoretical discussion of medical concepts and medical terms, it would probably be useful to concentrate interest on the practical usefulness of the concepts and terms in the time given, which in many cases would mean their relationship to the prevailing understanding of causation. This would at least be true with respect to preventive medicine, but probably to medicine in general too. To the doctor of the eighteenth century, a thorough description of the cough, the vomiting and the stools was his key to understanding, in the light of the theories for disease presented by the principles of humoral pathology; here he found his guidance for the proper treatment. To him, the diagnosis of "bilious fever" had an operative function based on causality considerations.

In modern preventive medicine, corresponding causality considerations, based on centuries of accumulated epidemiological and etiological knowledge, leads in many cases to conclusions where a medical concept covering the appropriate situation in the causality chain where intervention would most likely be successful — an operative concept of medical nature, when medical work should be extended into social, economical, political or private spheres — is simply failing.

7. DISCUSSION

Medical history and preventive medicine are disciplines closely connected, just because they look at causes and consequences of health and disease in perspective of time other than do workers in fields more directly related to individual patients. Therefore, work within medical history and preventive medicine inevitably gives birth to some suggestions for the future relating to the causality problem.

Diagnostics, viz the identification of medical entities requiring a certain set of reactions from the surroundings, is an extremely important part of medical work, especially when the diagnosis takes causation and consequences into account, thus representing an operative tool in the handling of the patient and in the prevention of similar cases in the future. On the other hand, diagnoses may in some cases appear rather as stumbling blocks to practical work or they may even seem to be mere collectors' items to the medical statistician. One should be sufficiently aware, in current revision of diagnostic practice, of the importance of the causality problem. The key is the assessment of the function of the diagnosis.

Because of the increasing influence of the concepts of health and disease prevailing outside the medical professions, the task of mediating medical information gradually acquires new dimensions — from a sheer enlightenment of the layman to the lobbying of a fellow player.

An important part of the informative work should be directed to the field of modern folk medicine, where the lack of understanding of the basic principles of scientific assessment of causality in medicine and biology provides fertile soil for erroneous concepts of causation in health and disease.

Probably a broader approach and a more extensive interdisciplinary orientation than has hitherto been used will be needed within medical information to place the concept of disease and the diagnosis, the medical entity, in a proper social context for the understanding of patients and society — to make the concepts, terms and interpretations properly operative when a problem has to be solved.

If we return for a moment to our example of the young smoker — to ask him to stop smoking would not even be the best prevention. The link in the causality chain most susceptible to intervention, probably lies elsewhere, prior to development of the habit observed. Research into the way of life in his subgroup of the population might reveal that his smoking only makes up a part of a set of habits and attitudes. A "sociological diagnosis" of these common customs of his subgroup and how these habits are acquired, would instead be the most suitable objective for preventive intervention.[5] The "operative concepts" needed in preventive medicine must be allowed a substantially wider sliding range on the scale between cause and consequence than is covered by ordinary diagnoses.

Progress in medicine and development of society have altered morbidity in a manner which is not always obvious to those not familiar with health statistics. Withdrawal of mortality to those age groups where life is apt to end anyway, makes mortality and its place in the medical causality chain of less practical value. Still, some statistics on occupational mortality are released, where conclusions were justified a generation ago, but rather cause confusion today. The increasing weight of "perceived morbidity" thus is not only a product of welfare, but a statistical reality which must be taken into consideration as such. Morbidity in a way takes over.

Lastly, it is a historical fact which cannot be overlooked that the medical professions of today are challenged by others, not only in the treatment of patients, but even more in the planning of health. The reasons for this are numerous; some of them can be traced back to the problems mentioned here.

Probably new needs will have to be met in the future — and these needs have to be foreseen by the medical schools in the first place. A deeper commitment to the problems of causality in medicine, the historical background and the social implications would mean a lot in the process of adjustment.

University of Oslo,
Norway

NOTES

[1] A more detailed outline of this topic is given in Larsen, 1981 [5]. Many of the ideas here emerged from fruitful discussions with Professor Arthur E. Imhof of Berlin and members of his group, which are duly acknowledged. Special reference is given to the introductory chapter of his book on biology in history ([1], pp. 13–78).

[2] The diagnoses have to be looked upon in the light of what they were used for. An

interpretation of the accuracy and specificity of a diagnostic term cannot be made without taking the contemporary medical setting into account.
[3] Literature on medical sociology and social medicine is abundant as to this point. We ourselves have presented a discussion of the "ecology of diseases" in [2].
[4] An epidemiological study of the changes in the concepts of disease as compared to the spread of infectious diseases in Norway in the period 1868 – 1900 gives some clues supporting this conclusion [4].
[5] Further discussion of this point is presented in our studies on the epidemiology of occupation and smoking habits [6].

BIBLIOGRAPHY

[1] Imhof, A. E. (ed.): 1976, *Biologie des Menschen in der Geschichte*, Frommann-Holzboog, Stuttgart-Bad Cannstatt.
[2] Imhof, A. E. and Larsen, Ø.: 1975, *Sozialgeschichte und Medizin*, Universitetsforlaget/Fischer, Oslo/Stuttgart.
[3] Larsen, Ø.: 1979, *Eighteenth-Century Diseases, Diagnostic Trends, and Mortality*, Department of Medical History, University of Oslo.
[4] Larsen, Ø., Haugtomt, H. and Platou, W.: 1980, *Sykdomsoppfatning og epidemiologi 1860–1900 (Infectious Diseases in Norway 1860–1900 and the Attitudes of the Health Authorities Towards Them – A Presentation of Data)*, Department of Medical History, University of Oslo.
[5] Larsen, Ø.: 1981, 'Separating Health and Illness – A Conceptual Framework', in: Gräsbeck, R. and Alström, T. (eds.), *Reference Values in Laboratory Medicine*, Wiley, London, pp. 33–43.
[6] Larsen, Ø., Leir, G. H., Wilhelmsen, D. S. S. and Aaløkken, T. M.: 1981, 'Røyking og yrke 1964 og 1980', *Norsk bedr.h.tj.* **2**, 204–224.

B. INGEMAR B. LINDAHL

COMMENTS ON LARSEN'S 'DISEASE FROM A HISTORICAL AND SOCIAL POINT OF VIEW'

Dr Larsen points out and discusses several conceptual and nosological pro-blems of importance to public health planning, and specially preventive medicine. I can only comment here upon a very limited part of this analysis.

Among the points I find particularly interesting to delve into a little deeper are the three fields Dr Larsen distinguishes where difficulties arise when problems of causality are handled in public health administration. Considering our limited time, however, I shall confine my comments to the five key notions in the first field, i.e., the different *faces* morbidity assumes, which I think are fruitful distinctions to be further developed. The notions or faces were: "true", "perceived", "influenced", "classified", and "admini-stered morbidity" ([1], p. 154ff.).

I should like to dwell briefly upon the definitions of these concepts, and examine to what extent they can be understood to exclude each other or have elements in common. I understand from the cloud allegory that these concepts represent different impressions or appearances of the same object. I conceive of this object as the morbidity of an individual or a population. I shall begin my commentary by suggesting some interpretations of these morbidity concepts in isolation. After that I shall discuss their possible relations to each other.

(*i*) *"True morbidity"*. "True" obviously does not mean here (a) "in accordance with fact or reality", which would have been an immediate lexical sense [2] (e.g. a morbidity estimation could be true in this sense). Dr Larsen seems rather to have in view (b) 'the uninfluenced disease conditions or morbid development itself', or (c) 'the unaffected impression of this phenomenon'. Although the "faces" are likened to impressions in the al-legory, the explication in terms of "the natural course and spread of a dis-ease" makes it clear that the morbidity itself is intended. I shall therefore choose (b) as my interpretation of "true morbidity".

(*ii*) *"Perceived morbidity"*. Dr Larsen describes this morbidity as the morbidity met with "through the complaints presented by the patients and the evaluation of the health problem in public opinion". I take this to mean that in most cases this is the mortality as we know it from patients' symp-toms and narrative sources etc., and not by physicians' direct observation of

165

L. Nordenfelt and B.I.B. Lindahl (eds.), Health, Disease, and Causal Explanations in Medicine, 165–167.

the patients' signs. I interpret "perceived morbidity" as an operational concept usually referring to those physical or mental phenomena which are labeled "morbid" by the patient himself or based only on the layman's evaluation.

(*iii*) "*Influenced morbidity*". Dr Larsen says that the true morbidity "may scarcely or never be observed". This morbidity is usually influenced by "the medical treatment and social handling of the particular condition at the time given". We can therefore rarely know what is "the natural course and spread of a disease". I understand this to be the essence of the concept "influenced morbidity". I interpret this concept as referring to the course or spread of the morbid physical or mental phenomena we usually see when we investigate diseased patients.

(*iv*) "*Classified morbidity*". The conceptual framework physicians use for diagnosis is itself influenced and constantly changed by the development and variations in "the prevailing professional principles for nosology". For example, two conditions *A* and *B* could be categorized as instances of the same kind of disease by a 19th century standard, and as examples of two different kinds of disease by another 20th-century standard. My interpretation of "classified morbidity" is simply that it refers to the morbid phenomena, categorized according to certain prevailing principles.

(*v*) "*Administered morbidity*" refers, as I understand it, to the fact that diseases are handled not only by direct medical treatment, but also indirectly through public health planning and policy making. The administered morbidity I interpret as the diseases in society as a whole, as influenced and managed by measures on a population level.

The true morbidity "may", as we quoted Dr Larsen earlier, "scarcely or never be observed". But *if* we observe some true morbidity it can, as I see it, be perceived (which is a way of observing), classified and administered. The true morbidity is in these cases not a separate kind of morbidity, but an example of perceived, classified and administered morbidity; the concept true morbidity does then to a certain extent overlap these three other concepts.

The notion perceived morbidity, it seems to me, clearly overlaps the influenced, classified and administered, since the perceived usually (maybe always) is influenced, and when the perceived morbidity is under medical care, it is also classified and administered.

The influenced morbidity is to a large extent not clinically observed, i.e., classified, or a case for health insurance etc., i.e., administered. Often diseases are taken care of by the diseased themselves, without clinical treatment or health insurance. But to some extent the influenced morbidity is also classified and administered and most (maybe all) of the classified and administered morbidity is influenced, according to my interpretation.

I further interpret Dr Larsen to mean that the classified morbidity is the basis for administered morbidity. In that case, these two concepts also overlap one another.

I have suggested here a few interpretations of the concepts of morbidity in Dr Larsen's text and far from all possible relations between them have been exhausted by this introductory commentary. My conclusion is that, on these interpretations, these notions are not all distinctly mutually exclusive. Some instances of diseases could be examples of more than one kind of morbidity in this system. A more systematic exploration of these concepts than we have been able to carry out here would, I am sure, be fruitful and of great interest.

Huddinge University Hospital,
Sweden

BIBLIOGRAPHY

[1] Larsen, Ø.: 1984, 'Disease from a Historical and Social Point of View: Some Remarks Based on the Needs of Preventive Medicine', in this volume, pp. 153–164.
[2] *The Concise Oxford Dictionary of Current English*: 1972, 5th Edition (1964), Oxford University Press, Oxford.

HENRIK R. WULFF

THE CAUSAL BASIS OF THE
CURRENT DISEASE CLASSIFICATION

1. INTRODUCTION

Most medical research workers take it for granted that it is one of the main purposes of medical research to explore the causes of different diseases. This is certainly true, but at the same time the statement is not quite as simple as it sounds. To illustrate this point I shall first use an example from daily life:

When a train arrives late we usually ask for *the cause* of the delay, but few people would ask for the cause when a train arrives on time. In fact, both late arrivals and arrivals on time are determined by a multitude of factors, and the reason why we ask for the cause in the former case but not in the latter, is simply that we do not like trains being late. We wish to find one factor which we can eliminate in order to prevent late arrivals in future. In just the same way, the state of illness − or the state of health − of an individual person is also determined by a multitude of factors, and whenever we seek *the cause* of the patient's illness we are in fact seeking a factor which we can eliminate in order to cure the patient. All events in and outside medicine are determined by numerous factors, and the selection of *the cause* is not a question of natural science; it depends on our interests which in medicine are often therapeutic or preventive.

Instead it may be regarded as the aim of natural science to explore the causal network as thoroughly as possible and to analyse the logical inter-relationship between the different factors, and for this purpose the concept of inus-factors, which was first introduced by Mackie, can be very helpful. I shall explain this idea by means of some clinical examples ([1], [2]).

2. CAUSES OF DISEASE IN THE INDIVIDUAL PATIENT

The first example concerns a man who develops a high temperature and neck stiffness. A lumbar puncture reveals that he suffers from meningitis, and the spinal fluid is found to contain pneumococci. Some time previously the patient had had his spleen removed after a traffic accident, and he was not vaccinated against pneumococcal infections.

169

L. Nordenfelt and B.I.B. Lindahl (eds.), Health, Disease, and Causal Explanations in Medicine, 169−177.

In this case we can list three causal factors:

$$\left.\begin{array}{l} \text{pneumococci} \\ \text{splenectomy} \\ \text{omission of vaccination} \end{array}\right\} \text{meningitis}$$

Most clinicians would undoubtedly say that the pneumococci constitute *the cause* of the infection, but that statement is debatable. The presence of pneumococci is not a *necessary* cause (as many patients with meningitis are infected with other bacteria or viruses), nor is it a *sufficient* cause (as the patient probably would not have developed the meningitis if he had not had his spleen removed or if he had been vaccinated). One might also argue that either the splenectomy or the omission of vaccination was the cause of the infection, but those two suggestions are equally debatable, as it also holds true of these factors that they constitute neither a necessary nor a sufficient condition for the development of meningitis. It is all three factors *in conjunction* which led to the patient's illness and, using the usual terminology, we may say that this complex of factors constituted a sufficient cause (as the patient developed meningitis) but not a necessary cause (as the patient could have developed meningitis in many other ways). The individual factors were neither necessary nor sufficient in relation to the development of meningitis, but each of them constituted a necessary component of this particular complex.

These logical considerations can be expressed quite briefly using Mackie's terminology: Each factor constitutes *an inus-condition*, i.e., an *in*sufficient but *n*ecessary part of an *u*nnecessary but *s*ufficient complex ([1], [2]).

In this case we have mentioned three inus-factors, and depending on our interests we may regard each of these as *the cause*. I said that most clinicians would regard the pneumococcus as the cause, and that is understandable as we may cure the patient by eliminating this factor by means of penicillin. However, the surgeon who removed the patient's spleen might feel a bit guilty, because he did not vaccinate the patient against pneumococcal infections, and he might feel that this omission was the real cause of the illness. He will then remember to vaccinate such patients in future in order to prevent repetitions of this unfortunate event. Finally, a theoretically minded immunologist who is more interested in immunological mechanisms than in clinical problems might hold that the absence of the spleen is the real cause.

I shall elaborate on these ideas by means of another clinical example.

A woman develops epigastric pain and we find that she suffers from a gastric ulcer. In a case like that we should usually say that the ulcer is *the cause* of the pain, but once again it is neither a necessary nor a sufficient causal factor. It is not a necessary one as patients may develop epigastric pain in many other ways, and it is not sufficient, as some ulcer patients have no pain. It is an inus-factor, i.e., a necessary component of a complex which also includes other unknown inus-factors, which I have labelled X, Y and Z:

$$\left.\begin{array}{l} \text{gastric ulcer} \\ X \\ Y \\ Z \end{array}\right\} \quad \text{epigastric pain}$$

The clinician decides to investigate this patient further and chooses to do a so-called pentagastrin test. He introduces a tube into the patient's stomach and finds, firstly, that her gastric acid production is quite normal and, secondly, that there is regurgitation of bile from the duodenum into the stomach.

The latter piece of information is interesting as we know that bile regurgitation may elicit ulcer formation. However, once again we have only found an inus-factor as not all persons with bile regurgitation develop an ulcer and as other factors, e.g., the ingestion of aspirin, may also cause ulcer formation. The logical interrelationship now looks like this, where A, B and C represent additional inus-conditions of gastric ulcer formation:

$$\left.\begin{array}{l} \text{bile regurgitation} \\ A \\ B \\ C \end{array}\right\} \left.\begin{array}{l} \text{gastric ulcer} \\ X \\ Y \\ Z \end{array}\right\} \quad \text{epigastric pain}$$

Then of course the clinician might study the motility of the pylorus in order to explain the regurgitation of bile, but for practical purposes it is not necessary to explore the causal network further back. The clinician will not try to treat the patient by eliminating the abnormal bile regurgitation, but he will do something quite different. I said before that the acid production was normal, which is quite common in gastric ulcer patients, but nevertheless the ulcer usually heals when the acid production is suppressed by drug therapy. In other words, the presence of acid is a necessary condition for the development of an ulcer, and it is possible to heal the ulcer by eliminating this normal but necessary factor. We shall insert the factor in the diagramme:

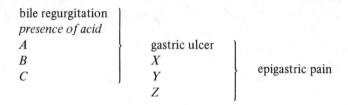

bile regurgitation
presence of acid
A   gastric ulcer
B   X   epigastric pain
C   Y
                           Z

I think that this example is illustrative as it shows that the delimitation of the sufficient causal complex also depends on our interests. Usually we select the causal factors among those factors which are abnormal or unusual, like the presence of pneumococci or bile regurgitation, but sometimes it is useful also to pay attention to those factors which occur normally.

I can give another clinical example of this phenomenon. A young girl suffers from heavy menstruations and develops an iron deficiency anaemia. In this case it would be natural to say that the loss of iron in connection with the menstruations was the cause of the anaemia, but in many cases the anaemia can be cured quite easily by giving the patient iron tablets. The iron intake with the food may have been quite normal, but logically speaking we may say that the "lack of an abnormally high iron intake" was one of the causes of anaemia, and we may cure the patient by eliminating this inus-factor.

I hope to have shown by these examples that Mackie's terminology is extremely useful for the analysis of clinical problems. All cases of illness are determined by a multitude of factors, and it is a Utopian thought that we should ever be able to map the whole causal network in any one patient. The total complex of factors which proved sufficient to cause the illness will always remain unknown, but the clinician may still be able to interrupt the disease process if only he can eliminate one necessary factor in that complex − if only he can eliminate a single inus-factor.

For the purpose of the arguments in this paper I need not take the analysis any further, but I may add that the course of the disease will depend on the position of the inus-factor in the causal network. We may for instance hope to cure the meningitis patient by killing the pneumococci as we are attacking the disease process at its start, but the girl whose anaemia was cured with iron tablets will become anaemic once again if we stop the treatment. Terms like symptomatic treatment, causal treatment, replacement therapy or preventive treatment all reflect different points of intervention in the causal network.

3. CAUSES OF DISEASE ENTITIES

In his analysis of causal relationships Mackie distinguishes between *the singular case* and *the general case*, and in just the same way we may distinguish between the *causes of illness in the individual patient* and *the causes of diseases*.

In everyday conversation we talk about diseases almost as if they were demons. We say that diseases attack people and we discuss their different manifestations. However, I shall only regard such *essentialistic* expressions as a mode of speech, and instead I shall adopt the *nominalistic* view that there are no diseases but only sick people. The disease classification is no more than a classification of patients, which serves to pidgeonhole clinical knowledge and experience and, consequently, the discussion of causes of diseases is no more than a generalization of the discussion of causes of illness in the individual patient [3].

In the 18th century very little was known about the causes of illness, and doctors could only classify their patients on the basis of symptoms and signs. Among "disease entities" from that time can be mentioned: phthisis (wasting), diabetes (passing through), typhus (stupor) and variola (mottled appearance). A few doctors like Carl von Linnaeus in Sweden and Boissier de Sauvages in France worked out elaborate taxonomies on the basis of all sorts of symptoms, but this idea was only short-lived. Nonetheless, we still base a number of diagnoses on the patients' clinical pictures, e.g. systemic lupus erythematosus and rheumatoid arthritis.

At the beginning of the 19th century the paradigm of the disease classification changed. Laënnec and many others developed the novel idea that diseases could be identified with anatomical lesions and numerous new disease entities were established.

Let us consider a single example. Doctors in the 19th century must have seen numerous patients who suffered from epigastric pain, and it was now appreciated from autopsy studies that some of these patients had a gastric ulcer. A new disease entity, gastric ulcer disease, was established. It has already been explained that the ulcer is only an inus-factor in relation to epigastric pain, but in accordance with the new paradigm the ulcer was regarded as *the cause* of the symptoms. It was also noted that some patients who died from a haematemesis also had a gastric ulcer in spite of the fact that they might have suffered no pain, and such patients were said to suffer from the same disease. It now became possible to write a new chapter in the textbooks with the heading "gastric ulcer disease" and the textbook author could describe which people the disease attacked, he could list the different manifestations of the disease and he could state its prognosis and its complications. The inus-factor, which had been selected as the cause because of the interest at that time in morbid anatomy, had suddenly attained the status of a disease demon, which can attack people and which can manifest itself in different ways.

Later that century, a new generation of medical scientists discovered bacteria and they, of course, established new disease entities, because in their eyes the bacteria were the real causes of disease. Previously, physicians could only distinguish between different clinical types of acute diarrhoea, but now cholera, typhoid fever, dysentery, etc. could be redefined according to the identity of the microorganism. This is an important phase in the history of the disease classification as it strengthened the false idea of the monocausal determination of diseases.

Mackie writes that yellow fever virus is a necessary cause of yellow fever and he believes that this statement is not tautologous as the virus and the disease can be defined separately. I think that he is wrong on this point. There are mild cases of infection with yellow fever virus which no clinician could recognize clinically with certainty, but we should still say that the patient suffered from yellow fever, if we succeeded in isolating the virus from a blood sample. Similarly we can imagine clinical pictures, which are indistinguishable from typical yellow fever, but we would not say that such patients suffered from yellow fever, if some other infective agent was isolated. Yellow fever has been redefined to mean yellow fever virus infection. I believe that this example is illustrative and I shall suggest that causal factors are never (or very rarely) necessary in relation to a disease entity except by definition. I shall explain later that it is one of the main purposes of the disease classification to establish classes of patients who have a necessary factor in common.

Towards the end of the 19th century and in this century doctors developed an interest in physiology and biochemistry, and a multitude of new inus-factors were recognized. This development of course led to new disease entities, like for instance hyperthyroidism, porphyria and diabetes mellitus (defined by the disturbance in carbohydrate metabolism). Nowadays immunological research has attracted a lot of interest, and this epoch in the history of medicine will undoubtedly lead to immunologically defined disease entities.

I shall summarize my arguments briefly. Illness in the individual patient is always determined by a multitude of factors. Doctors sometimes succeed in discovering some of these, and they usually select one, which they label *the cause* of the illness in spite of the fact that it may only be an inus-factor. This selection is greatly influenced by the ever changing traditions in the history of medicine. Sometimes doctors succeed in recognizing the same causal factor in a number of their patients, and they are then liable to create a new disease entity. However, we usually forget that the disease classification is our own creation, and instead we talk about *the discovery* of new diseases as if they had been there all the time.

4. THE PURPOSES OF THE DISEASE CLASSIFICATION

I believe that it is very important that doctors realize that the disease classification is no more than a classification of patients which reflects consecutive medical traditions, because if they ignore this fact they will not ask themselves to what extent the classification serves different purposes. Is it equally well suited for hospital practice and for general practice? Is it equally well suited for therapeutic and preventive purposes?

In this paper I shall only raise one of these points and I shall suggest that the interests served by the current classification are mainly therapeutic. The causal factors which determine the development of illness can be divided into aetiological and pathogenetic factors. The aetiological factors are those which elicit the disease process and they can be either genetic or environmental, e.g., what we eat, what we inhale, which microorganisms surround us and in general how we live. The pathogenetic factors on the other hand pertain to the disease process inside the human body, like anatomical lesions or physiological disturbances. This distinction is very important because knowledge of environmental (aetiological) and pathogenetic factors serve different purposes. If we succeed in isolating pathogenetic inus-factors which we can eliminate, we may be able to cure our patients or perhaps to stop the progress of the disease in that particular patient, but of course that knowledge does not permit us to prevent the disease. If, on the other hand, we isolate an environmental factor, we may be able to prevent the development of disease, but we may not be able to cure our patient if the disease process has started. This is the general rule, but there are exceptions. We may, for instance, cure an asthma patient by removing the allergen from the enironment, and we may prevent infectious diseases by vaccination.

This line of thought also reveals the purpose of diagnosis. The present disease classification is to a large extent based on pathogenetic inus-factors, and very often clinical diagnosis amounts to the demonstration of a factor which is a necessary part of the complex which led to the disease in that particular patient. If this factor is eliminated the patient may be cured. The diagnosis of hyperthyroidism tells us to give antithyroid drugs, the diagnosis of B_{12}-deficiency anaemia tells us to start treatment with vitamin B_{12}, and the diagnosis of a cancer of the colon suggests surgical treatment. Clinical syndromes, like systemic lupus erythematosus or rheumatoid arthritis are less satisfactory from a practical point of view, as these diagnoses are not causally defined but only reflect the patients' clinical pictures. It must be hoped that medical scientists will succeed in elucidating the causal mechanisms in such patients so that it will prove possible in future to reclassify the patients according to the presence of particular causal factors.

It also follows from this discussion that the present disease classification is less suited for preventive purposes. The diagnosis of a disease like iron-deficiency anaemia is clinically helpful as it suggests which further investigations the clinician should do and as it suggests a specific treatment, but at the same time it is almost useless for preventive purposes. Some patients with this disease eat too little iron, others do not absorb iron from the gut, and still others suffer heavy menstruations or bleed from, for instance, a duodenal ulcer. It is an impossible task to find one preventive measure which can "eradicate" this particular disease entity.

A disease like myocardial infarction probably presents a similar dilemma. The factor which the patients have in common is the anatomical lesion, but we must imagine that the combination of factors which started off the process differs from patient to patient. Epidemiologists have done their very best to unravel the complex aetiology, but the complexity of the problem is shown by the fact that they always end up finding a number of risk factors which may well be statistically significant, but which do not help us to prevent the disease with any degree of certainty in the individual person. I appreciate that we have been able to prevent many infectious diseases, but my guess is that in most other cases it will prove impossible to find a single environmental inus-factor, which will permit effective prevention of a particular disease.

This does not mean that I am not in favour of preventive medicine. I agree that prevention is very important, but we may be attacking the problem in the wrong way. Perhaps we should not spend so much time reasoning backwards from effect to cause, as it might prove more profitable to do prospective studies of the effects of different environmental factors. We might for instance follow cohorts of people who are exposed to certain chemicals, who smoke too many cigarettes, who are unemployed or who are living under career pressure, and we might well find that one aetiological factor may contribute to the development of a lot of illness, which cannot be squeezed into the therapeutically orientated disease classification.

These methodological considerations also suggest that it might be profitable to approach the logical analysis of causal interrelations from a different angle. It is typical of discussions on the logics of causality that the direction of reasoning is from effect to cause: an effect is specified and it is discussed whether a number of factors are necessary, sufficient or (like inus-factors) neither necessary nor sufficient. I hope to have shown that this type of analysis fits clinical reasoning which is therapeutically orientated, but it does not fit the type of reasoning which is needed in preventive medicine. For that purpose it would be more reasonable to specify a causal factor and to analyse the logical status of a number of effects. A "mirror image" of the conventional terminology would be needed and one might, for instance, distinguish between obligatory, possible and impossible effects.

Analysis from effect to causal factors (from disease to causes of disease) also has the disadvantage that the idea of disease is taken for granted, whereas analysis from cause to effects invites a distinction between disease-promoting and health-promoting effects. In that way the discussion of causality in medicine would be linked to the discussion of the concepts of disease and health.

Herlev University Hospital,
Denmark

BIBLIOGRAPHY

[1] Mackie, J. L.: 1965, 'Causes and Conditions', *American Philosophical Quarterly* **2**, 245–64.
[2] Mackie, J. L.: 1974, *The Cement of the Universe: A Study of Causation*, Clarendon Press, Oxford, U.K.
[3] Wulff, H. R.: 1981, *Rational Diagnosis and Treatment. An Introduction to Clinical Decision Making*, 2nd ed., Blackwell, Oxford, U.K.

H. TRISTRAM ENGELHARDT, JR.

COMMENTS ON WULFF'S 'THE CAUSAL BASIS
OF THE CURRENT DISEASE CLASSIFICATION'

Causal explanations function in medicine, as Henrik Wulff shows in his paper, within the embrace of a number of quite pragmatic considerations. Since medicine is not a pure science, but an applied science, it is not so much concerned with fashioning true classifications or true diagnoses, in some contextless sense of truth [1]. It is rather focused on maximizing well-being and minimizing the suffering of patients or possible patients. Since the probabilities of most diagnostic choices and the therapies they warrant being the most appropriate lie somewhere between zero and one, physicians must frame accounts of reality in terms of the alternative costs of different ways of being wrong. Which is to say, in framing a diagnosis, or giving a causal account of a particular clinical problem, a physician must assess the costs and benefits of acting as if the probability value of that choice being a correct choice were one, though realizing that it is not the only one. One might recall here C. S. Pierce's definition of a pragmatic account of truth: "In order to ascertain the meaning of an intellectual conception one should consider what practical consequences might conceivably result by necessity from the truth of that conception; and the sum of these consequences will constitute the entire meaning of the conception" ([4], 5.9). With an appropriate specification of "practical" consequences one can deliver the instrumentalist view of truth which Henrik Wulff uses in accounting for diseases and causes of diseases [8].

With respect to causes, Wulff indicates how medicine picks certain causal variables from a field of causal factors. As he puts it, "the selection of *the cause* is not a question of natural science; it depends on our interests which in medicine are often therapeutic or preventive" ([8], p. 169). In this he is quite correct. If one were to take exception to his presentation of causal accounts in medicine, it would be to accent further the difference between such accounts as they occur in the applied sciences, versus those available in unapplied sciences. Applied sciences are directed towards identifying some causal variables out of a field of causal factors, because it is useful to hold those accountable for the outcomes in question. Usefulness here is usually tied to some practice of intervention or assigning responsibility. Physicians must select those causes which are usually treatable. Often, causes will include omissions ranging from the failure to acquire proper immunizations,

179

L. Nordenfelt and B.I.B. Lindahl (eds.), Health, Disease, and Causal Explanations in Medicine, 179–182.
© 1984 by D. Reidel Publishing Company.

to the failure to engage in regular exercise. In this respect, as Kenneth Schaffner has indicated, causal language in medicine has great similarities to the ways in which causes are identified in the law [5].

Here the very influential study, *Causation and the Law*, by Hart and Honoré must be acknowledged [3]. Hart and Honoré outlined how the law identifies causes by appeals to accepted practices. Against the background of an accepted practice on can identify which departures ought to be held responsible for an outcome. As a result, the law often identifies as causes of an outcome omissions rather than commissions. One is led to identifying a particular factor as *the cause*, because of interests in assigning responsibility or accountability. So too, medicine selects particular causal variables out of interests in framing therapeutic interventions or providing predictions of use to physicians and patients [2]. Casual factors, which cannot usefully be construed in this fashion, are treated as background conditions. Thus, in the case of infectious diseases, though there may be genetic variability in the extent to which particular humans are susceptible, such genetic factors are usually not of much interest to clinicians insofar as effective antibiotics are available. Differences in genetic resistance are shoved into the background. This suggests that physicians will vary in their identification of causes, depending upon their specialties, and in the effective approaches to therapy their specialties allow. Thus, a medical geneticist, a surgeon, and a psychiatrist may very well accent different causes in giving an account of the development of heart disease or duodenal ulcer.

Here it should be clear that I am in agreement with Wulff's account and am in fact only suggesting a further development of what I take to be a part of his analysis. If there is a shortcoming, it lies in his not having as fully underscored, as he might have, the special instrumentalist interests that are a part of applying causal explanations within a practice that is concerned with maximizing particular outcomes other than those of general powers of prediction and explanation. Wulff is correct in tying medicine's causal accounts to classifications of disease. Since classifications of disease in clinical medicine function as indices to treatment warrants, it is appropriate for them to identify useful causal factors. That is, they must point out variables that can be manipulated so as to remove the disease, prevent the disease, or ameliorate some of its consequences. Which is to say, when causal language is employed in medical decision making, it is primarily dependent for its usefulness, and therefore for its meaning, on the ways in which it indicates manipulable causal factors, though it also plays a role in offering patients prognoses.

In his assessment of the etiological dimensions of so-called disease entities, Wulff contributes to the distinction of clinical accounts from classical accounts of disease entities. In addition, he joins a rather long history, critical of the conflation of etiological factors with *the* disease. One might think here of Rudolf Virchow's rather famous distinction between an *ens morbi* and a *causa morbi*:

An actual parasite, whether it is plant or animal, can become the cause of a disease, but it can never exhibit the disease itself. Nothing demonstrates the necessity of this differentiation more than the fact that parasites in great numbers live even in the healthy body, and that in individual cases parasites can be harmless, even though they can act as pathogens (more strictly, pathogenically) according to current usage. If this action does not take place, if, e.g., diphtheria bacilli are in the throat of a healthy child, then there is no disease and therefore no *ens morbi* present ([6], p. 120).

In addition to distinguishing causes of disease from disease entities, Virchow is offering the beginning of a distinction between necessary and sufficient casual factors in the development of accounts of disease. An approach such as Wulff's adds the further refinement of pointing out that there will not be one unambiguous sense of *the* cause. Rather, one must accent particular casual factors in particular kinds of diagnostic, and therefore eventually therapeutic, decisions. Indeed, mapping *all* the causal factors would not as such, even if possible, be of use to clinical medicine. One would also need to know which causes it is most cost-effective to address and when.

Through using Mackie's account of causality, Wulff is able to display the richness of the causal field within which physicians must act. However, it would have been useful to have indicated, as Kenneth Schaffner has pointed out, ". . . that one must alter Mackie's appeal to 'elliptical or gappy universal' propositions to read 'general' propositions in order to allow the use of statistical generalizations" ([5], p. 120). Physicians are, as Wulff acknowledges, making various statistically based bets. The causal variables that physicians must identify are, as a result, often not strictly necessary parts of unnecessary but sufficient causal complexes, but rather marks in favor of the greater likelihood of a particular diagnosis being correct. This emendation is required in order to capture the pragmatic character of the selection of causes. One is attempting to find indications that will allow one to make a choice of a cause in a way that will be most cost-effective with regard to interests in avoiding morbidity, risks of mortality, as well as financial costs.

The criticisms that I have offered are fully in accord with the general premises of Wulff's arguments, both in this paper as well as in those he has developed in his book, *Rational Diagnosis and Treatment* [7]. This approach offers an important step towards appreciating medicine as clinical medicine, as an applied science. That is, it is an applied science in the sense of employing empirical generalizations, not in order to offer explanations or predictions as ends in themselves, but in order to apply them towards the realization of non-epistemic goals, here those such as the avoidance of certain impairments. One must suspect that much of the misunderstanding in the history of medicine regarding the nature of causes is dependent upon a view of medicine as an unapplied science. This view has obscured the instrumentalist interests in truth that clinicians possess. Wulff has made an important contribution to the reappraisal of our understanding of causal and disease explanations in medicine, and to the appreciation of clinical medicine in particular.

Baylor College of Medicine,
Houston, Texas

BIBLIOGRAPHY

[1] Engelhardt, H. T., Jr.: 1975, 'The Concepts of Health and Disease', in S. F. Spicker, and H. T. Engelhardt, Jr. (eds.), *Evaluation and Explanation in the Biomedical Sciences*, D. Reidel Publishing Company, Dordrecht, pp. 125–142.
[2] Engelhardt, H. T., Jr.: 1981, 'Relevant Causes: Their Designation in Medicine and Law', in S. F. Spicker *et al.* (eds.), *The Law-Medicine Relation: A Philosophical Exploration*, D. Reidel Publishing Company, Dordrecht, pp. 123–128.
[3] Hart, H. L. A., and A. M. Honoré: 1959, *Causation and the Law*, Clarendon Press, Oxford.
[4] Peirce, C. S.: 1965, *Collected Papers*, C. Hartshorne and P. Weiss (eds.), Belknap Press, Cambridge, Mass.
[5] Schaffner, K. F.: 1981, 'Causation and Responsibility: Medicine, Science and the Law', in S. F. Spicker *et al.* (eds.), *The Law-Medicine Relation: A Philosophical Exploration*, D. Reidel Publishing Company, Dordrecht, pp. 95–122.
[6] Virchow, R.: 1981, 'One Hundred Years of Clinical Pathology', S. G. M. Engelhardt (trans.), in A. Caplan *et al.* (eds.), *Concepts of Health and Disease*, Addison-Wesley, Reading, Mass.
[7] Wulff, H.: 1981, *Rational Diagnosis and Treatment*, Blackwell Scientific Publishers, Oxford.
[8] Wulff, H.: 1984, 'The Causal Basis of the Current Disease Classification', in this volume, pp. 169–177.

GERMUND HESSLOW

WHAT IS A GENETIC DISEASE?
ON THE RELATIVE IMPORTANCE OF CAUSES

1. INTRODUCTION

We generally think of a genetic disease as a disease that is genetically caused or as a disease for which the genetic cause is more important than environmental causes. Diseases like PKU (phenylketonuria) and lactose intolerance, for instance, are classified as genetic, because they are caused by enzyme deficiencies which are due to genetic defects. In both cases the absence of a gene (or the absence of a gene with an adequate structure) leads to the absence of an enzyme that is able to counteract a harmful substance. The substance, phenylalanine or lactose, accumulates and damages the organism.

But this is not enough to justify our calling these diseases genetic. With a perfectly analogous argument we could say of someone who has just died from arsenic poisoning, that the lack of an enzyme that counteracts arsenic led to accumulation of arsenic and to the death of the victim. Since the lack of such an enzyme is due to the absence of an adequate gene, the victim suffered from a genetic disease. This is absurd, of course — but why? Why are we so reluctant to recognize arsenic poisoning as a genetic disease?

Although it is not without intrinsic interest, this question is important mainly because it illustrates a much wider and more fundamental issue, namely that indicated in the subtitle of this paper "the relative importance of causes". It is a generally recognized fact that most events will have a large number of causes, not only because of the possibility of tracing a causal chain backwards in time, but also because there will be many events at any particular point in time which interact to give a certain effect. Most of these events are never cited in causal explanations, however. We generally select a few, often just one, of them as the most important or the "decisive" causes. Sometimes the relative importance of causes is graded, as when we say that genetic factors are more important than environmental ones although the latter contribute.

This gives rise to the two questions with which we shall be concerned here. Firstly, given a situation in which several factors jointly cause a certain outcome, is it meaningful to select one of them as "the" cause or as the "most important" cause, or are such selections arbitrary? Secondly, if it is

183

L. Nordenfelt and B.I.B. Lindahl (eds.), Health, Disease, and Causal Explanations in Medicine, 183–193.

meaningful, what does it mean and which are the criteria that govern such selections and assignments of different weight to different causal conditions?

As testified by the numerous references to it during these proceedings, this problem is relevant to several other issues in the philosophy of medicine. As noted in the introductory paragraphs, causal weightings determine the classification of diseases into broad categories like genetic versus environmental and somatic versus psychosomatic diseases. Classification of diseases into specific "entities" is also based on the causes of certain clusters of symptoms and will to some extent reflect judgements about the relative importance of various causal conditions. As shown for instance by Dr Fagot ([2], pp. 102–107), selections and weightings of causes enter into assignments of responsibility and legal liability.

There is, finally, an important heuristic aspect of all this. There are certain kinds of intellectual patterns, "paradigms" if you wish, which influence our ways of thinking and decision making. These can often be difficult to formulate, because once they are made explicit people easily recognize their falsity, thus making the point of exposing them look trivial. Let us nevertheless consider two examples of such heuristic principles:

(a)   If C_1 is a more important cause of E than is C_2, then manipulating C_1 is a more effective way of influencing E than is manipulating C_2.

When it is said, for instance, that 90% of all cancers are environmentally caused, the implication is often that the best, or even the only, way of preventing cancer is to change the environment.

(b)   If C_1 is a more important cause of E than is C_2, then, in studying the mechanisms by which E comes about, it will usually be more profitable to concentrate on C_1.

For instance, if it is known that a certain disease is environmentally caused, many scientists will collect data pertinent to dietary factors or industrial pollutants, whereas the knowledge that there is a genetic cause will increase the frequency of researchers looking for chromosomal aberrations.

Note that I am not saying that many people actually believe in either of these principles, but only that they are biased towards them in their thinking, and that the reason for this might be a conceptual confusion about the real meaning of causal selections.

2. THE RELATIVE IMPORTANCE OF CAUSES

My basic point can be made very simply. Suppose that we have three conditions which are individually necessary and jointly sufficient for the effect E.

$$\left.\begin{array}{c} C_1 \\ C_2 \\ C_3 \end{array}\right\} \rightarrow E$$

Now if this was *all* we had, I would be inclined to think that any selection from these conditions would be quite arbitrary, and any claim that one condition was more important than the others would be meaningless. But suppose that we have another case, another individual, where only two of the three conditions are present and, consequently, where the effect is absent.

$$\left.\begin{array}{c} \sim C_1 \\ C_2 \\ C_3 \end{array}\right\} \rightarrow \sim E$$

Suppose now that we are not asking why E occurred in the first case, but instead why there was a *difference* between the cases, i.e., why E occurred in the first case, *when it did not in the second*. When the question is rephrased in this way, C_1 takes on a special significance because C_1 is the *only* condition that can explain the difference. C_2 and C_3 are present in both cases and are therefore incapable of explaining the difference between them.

There is, clearly, nothing arbitrary about the claim that C_1 is the most important condition of the three, but the example illustrates an important proviso. The selection is meaningful only if the question that the causal statement is supposed to answer concerns a difference between different occurrences of the effect and if the choice of some specific occurrence as an object of comparison is not arbitrary.

Consider the stylized fruit flies in Figure 1, and let us first compare those that are designated $M\,1$ and $N\,1$. $N\,1$ is a genetically normal fly, and $M\,1$ is a mutant. It is assumed that they have been brought up under identical environmental conditions. If we now put the question "Why does $M\,1$ have such short wings?", the answer seems obvious. The cause is genetic — *$M\,1$ is a mutant*. But suppose that we had never seen $N\,1$, and that $M\,1$ had only been observed together with the flies $M\,2$ and $M\,3$, both of which are genetically identical to $M\,1$ but have been raised in higher temperatures (27° C and

Mutation   Normal

22°

M 1

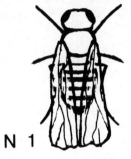

N 1

27°

M 2

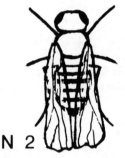

N 2

32°

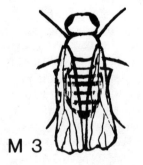

M 3

N 3

Fig. 1.

32°C respectively, as compared to 22°C for *M* 1). If we look at these and put the question again, "Why does *M* 1 have such short wings?", the answer will again be obvious, although it will now be a different one. The wings are short because *M 1 was raised at low temperature.*

When trying to explain a property like the short wings of a fly, we seem to be making a comparison between this fly and a different one. We may be unaware of this comparison, but it is the most natural way of making sense of the fact that the answer to our question is dependent on, and varies with, the object of comparison.

It should be clear that conditionship relations are quite irrelevant here. In a wide and rather trivial sense we could of course say that the genes are important for the length of the wings. For if the genes *had* been different from what they are, the wings *could* have been different. And if all the flies had been raised at room temperature, we could still say that the environment had something to do with the wings, for if the temperature *had* been higher, the wings *would* have been larger. But in this sense arsenic poisoning would be a genetic disease, for if our genes had been different from what they are, we *might* have been able to produce something that would counteract arsenic.

In this wide sense every disease is caused by a combination of genes and environment and no weighting is possible. The relative importance that we ascribe to these factors in many cases is determined by the extent to which they are able to account for the difference between the individual under consideration and those which are selected as objects of comparison. If we compare an individual with others that are genetically different and have shared the same environment (as *M* 1 and *N* 1), then the genes will be the more important, because only the genes can account for the phenotypic difference. Comparing with genetically identical individuals which have had different environments (as *M* 1 and *M* 2) will make the environmental factor the more important one.

In general we are not explicit about which the objects of comparison are — we are usually not even aware that there are any — and the reason for this is simple. We generally compare with some "ideal type" in the back of our minds, and since this ideal type is formed by experience, it is likely to be composed of traits which are "normal" in a statistical sense. Had all fruit flies been of the kind in the first row, we would have tacitly compared *M* 1 with *N* 1 and the genes would have been decisive. Had all flies been of the kind in the first column, we would have compared *M* 1 with *M* 2 and *M* 3 and the environment would have seemed most important. This is why arsenic poisoning is not a genetic disease. The lack of an arsenic counteracting enzyme does not account for the difference between people who are poisoned and others, simply because there are no others with whom it would be natural to compare, namely people who have such enzymes.

Although the objects of comparison will often simply be those which are normal or typical, it is important to realize that they do not have to be. Even if only a few flies had been raised in higher temperatures, it would make perfectly good sense to compare $M\,1$ with $M\,2$ and $M\,3$, provided that this choice is made explicit. Being explicit about the objects of comparison might not be a bad idea anyway, for "normal" is a highly ambiguous term[1] and will mean different things to different people. What we conceive as normal will to some extent depend on personal experience and subjective factors like personal interests and point of view. But if we are careful in formulating our questions, making it clear precisely which difference is involved in the explanandum, subjective bias will have no leeway.

The view of causal selection which has been sketched above is consistent with most of the alternative suggestions which have been proposed in the literature. Indeed, it treats them as reducible to what I believe to be the more fundamental principle and explains them as reflecting different choices of the objects of comparison. It is only possible to exemplify this very briefly here.[2]

(a) Perhaps the most popular view is that those conditions which are picked out are those which are abnormal. This, of course, is equivalent to picking out those conditions which explain why an individual differs from what is normal.

(b) We sometimes pick out "precipitating" conditions: the spark that ignites the explosion, etc. This is equivalent to explaining the difference between this individual or object now and at an earlier time.

(c) Some philosophers have argued that it is impossible to separate assignments of causality from those of blame, because we sometimes pick out as the most important causes those conditions which *should* not have occurred. This is equivalent to making a comparison, not with what is normal, but with what is "appropriate".

(d) It is common, particularly in theoretically advanced sciences like physics and economics, to use as objects of comparison purely hypothetical states, idealizations which are defined by abstract theories. It is not explained in classical physics, for instance, why an object moves exactly as it does, but why it deviates from a state of "uniform motion". An economist may similarly explain why the economy deviates from a model of perfect competition. In medicine this is still rare, though the concept of physiological health proposed by Boorse [1] may be looked upon as an attempt to provide medicine with such an idealization. The advantage of this practice is that the object of comparison will be permanent and not subject to context-dependent variations.

3. THE RELATIVITY OF CAUSAL IMPORTANCE

Accumulation of lactose in the intestine causes diarrhea and abdominal cramps, and this will occur if, firstly, the patient consumes milk or milk-products and secondly, he does not have a certain amount of lactase. In northern Europe most adults produce a sufficient amount of lactase, and milk consumption will seldom lead to any trouble. If each column represents a Scandinavian individual, the situation can be visualized in the following manner:

$$L\,L\,L\,L\,L-L\,L\,L\,L\,L$$
$$MMMMMMMMMMM$$
$$\downarrow$$
$$S$$

L here stands for the presence of lactase, M for regular milk-consumption and S for the symptoms diarrhea and abdominal cramps. The symptoms are present in only one individual, the same one who lacks lactase, while everyone consumes milk. Since only the difference in L is reflected in the difference in symptoms, we would ordinarily choose the absence of L as the most important cause of S. Since the absence of L is genetic, S is a genetic disease.

But consider now an African who suffers from S. Lactase deficiency is a pretty normal condition among African adults, while milk-consumption is rare. The situation is this:

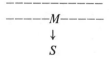

The only factor that differentiates the patient from a normal person is M, and since this is an environmental factor, S is an environmental disease.

The symptoms are caused in both cases by a combination of milk and absence of lactase, but since the natural objects of comparison vary, the relative importance of genes and environment also varies.

The relativity in the previous case was geographical. A different example will illustrate a similar temporal relativity. Table I gives the probabilities that relatives of someone with a certain disease X will also be affected by X.

TABLE I
Percentage frequencies of X in the families of persons with X.
From Stern [6].

Relationship to affected person	Percentage affected
Unrelated general population	1.4
Spouses	7.1
Parents	16.9
Half-sibs	11.9
Sibs	25.5
Nonidentical twin	25.6
Identical twin	87.3

These figures are fairly typical of those diseases we usually call genetic or inherited. The correlation between spouses is small, although it does point to some environmental influence. The probability of contracting X increases from 1.4% to 7.1% if you marry someone with X, and live in a similar environment. But if you also share the genes of someone with X, as an identical twin does, the probability increases to 87.3%. These figures clearly point to a strong genetic influence, slightly stronger in fact than for diabetes, a disease which is sometimes classified as genetic (e.g., [5]). It may be somewhat surprising, therefore, to find that X is actually an infectious disease, namely tuberculosis, although the figures are some forty years old.

Forty years ago almost everyone came in contact with the tubercle bacillus, and the factor that decided whether someone was to develop the disease was therefore not the exposure to the bacillus but the individual susceptibility. When everyone has been exposed the bacillus, this cannot explain the differences between individuals: why x got tuberculosis when y did not. Since contraction of the disease coincided with susceptibility, this factor, which is genetic, can explain such differences.

The high weight given to the genetic factor in this example is a result of taking normal cases as the objects of comparison, normal cases having been exposed to tubercle bacilli. This weighting could be different if the question were asked differently. The genes will dominate if we ask why x got the disease, when y and z did not. But if we ask instead why x got tuberculosis this year, when he was perfectly healthy before, the environment may have been decisive.

In the industrialized world today tuberculosis is a fairly rare disease, and only a very small part of the population will ever come in contact with the bacillus. Furthermore, since most people are vaccinated, susceptibility is no longer entirely genetic. The combined effect of this is that tuberculosis patients are no longer differentiated from the rest of the population by their higher susceptibility (Table I would look quite different today), and the disease has largely ceased to be a genetic one.

4. CONCLUSIONS

I want to end this paper with some methodological reflections prompted by the previous argument.

Causal strength. When a cause of some particular phenomenon has been judged more important than some other condition, it is very tempting to regard this as being somehow inherent in the causal relation itself, as reflecting the "strength" or "effectiveness" with which the cause brings about the effect. It should be clear by now that this is not so. The fact that genes in a particular case are more important than the environment is not primarily a fact about the disease itself, or about the causal mechanism, or about the strength of the causal association. It is mainly a fact about the relative frequencies of the possible objects of comparison. Lactose intolerance (in Scandinavia) is not genetic because the genes exert a stronger influence on the organism, but because lactase deficiency is rare and milk-consumption common. Since the frequencies of objects of comparison will vary in geographical, temporal and other dimensions, the relative importance of causes will vary correspondingly.

Causal classification. If the causation of an effect, a disease for instance, were an absolute fact only about this event, then it might be natural to regard the class of similarly caused effects as "real". A disease classification based on "most important" causes would reflect an objective reality, a natural order among things as contrasted with an order imposed by the intellect. The fact that what are the "most important" causes is not something intrinsic to the events entering into the causal relation, but is relative and subject to variation, considerably weakens this realist stance. It would be going to far to say that the concept of a genetic disease is nonsensical, but it certainly makes much less sense than is widely believed.

Manipulability. It is a view implicit in much medical thinking that knowledge about the relative importance of causes is important for the possibilities of fighting a disease. Knowing that the genes are important would imply for instance, that the disease is correspondingly more difficult to affect, because genes are impossible to manipulate, and you must manipulate the cause in order to influence the effect. This view is clearly mistaken. To prevent an effect we certainly have to prevent *some* cause, but any cause will do, and the fact that one cause is more important than another has nothing to do with its instrumental efficacy.

Scientific strategy. As mentioned in the introduction, beliefs about the importance of broad classes of causes such as genes and environment tend to direct scientists in their search for the finer details of causal mechanisms as if these details were somehow unavoidably to be found within the most important of the broader classes. As a general heuristic principle such a strategy may be useful, but it is not without its dangers. Most of us would, when forced to think about it, agree that a strong hereditary pattern like the one illustrated by tuberculosis does not preclude the possibility that a disease could be cured by the elimination of an environmental agent. But, in all honesty, I doubt that many readers, except perhaps specialists in this particular field, when looking at Table I gave much thought to the possibility that X might be an infectious disease. I think, therefore, that a better understanding of what is involved in the weighting of causes should make it easier for us to realize and take account of possibilities which we might otherwise have overlooked.

Preciseness of questions. The final and, though somewhat simplistic, most important point is that the relative weight that we ascribe to different factors is determined by the questions we ask, and the argument thus emphasizes the importance of being clear and precise in formulating questions. As we saw in the fruit fly example, the key to the problem of causal selection lies in the choice of objects of comparison. Since we are usually unaware that we are making a comparison and that our judgement in a specific case is dependent on a context taken for granted, we seldom see the need to clarify our problems in this way. But there are several examples of scientific controversies in which at least part of the problem seems to be a conceptual confusion about the nature of the problem rather than a lack of sufficient data. The causes of mental illness and alcoholism are cases in point.

The main cause of the latter, for instance, will be different as answers to the following questions: (a) Why is Smith an alcoholic, when Jones is not? (b) Why is Smith an alcoholic when his twin brother is not? (c) Why did Smith suddenly take to the bottle, when he used to be so careful with alcohol? (d) Why are so many Westerners alcoholics when so few Arabs are? If Smith and Jones are both unemployed, unemployment cannot explain the difference between them but it might explain the difference between Smith and someone else. Genes cannot explain the difference between Smith and his twin brother, but unemployment might. The best answer to (d) would probably be the availability of alcohol, but this is not a plausible answer to any of the other questions.

The moral is obvious and it reinforces the often repeated but seldom heeded rule that nothing is so important in intellectual work as being clear and precise about the problem one is trying to solve.

University of Lund,
Sweden

NOTES

[1] A nice discussion of the various meanings of "normal" is to be found in Murphy [4].
[2] A number of the criteria of causal selection and weighting proposed in the literature are discussed in more detail in my [3]. This paper also contains a more technical exposition of the basic ideas of the present paper.

BIBLIOGRAPHY

[1] Boorse, C.: 1977, 'Health as a Theoretical Concept', *Philosophy of Science* **44**, 542–73.
[2] Fagot, A.: 1984, 'About Causation in Medicine', in this volume, pp. 101–126.
[3] Hesslow, G.: 1983, 'Explaining Differences and Weighting Causes', *Theoria*, in press.
[4] Murphy, E.: 1976, *The Logic of Medicine*, Johns Hopkins Press, Baltimore, Maryland.
[5] Robbins, S. L. and Angell, M.: 1977, *Basic Pathology*, W. B. Saunders Company, Philadelphia.
[6] Stern, C.: 1973, *Principles of Human Genetics*, Freeman, San Fransisco, California.

HENRIK R. WULFF

COMMENTS ON HESSLOW'S
'WHAT IS A GENETIC DISEASE?'

I agree with Hesslow in what I take to be his main conclusion: diseases are determined by both genetic and environmental factors, and the importance which we attach to different factors depends on our choice of causal field. This is an important message, as the clinician who stares himself blind at the genetic factor may well overlook that it is possible to prevent or cure the disease by the elimination of a necessary environmental factor. However, Hesslow may be going a bit too far. The paper is entitled 'What is a genetic disease?', and the reader is left with the impression that it is simply a matter of choice whether or not a disease is labelled as genetic. In this comment I shall suggest what I consider to be reasonable definitions of the concept of genetic disease, and I shall discuss a number of examples. First, we should consider the situation where *a genetic abnormality is a necessary and a sufficient determinant of a disease regardless of the environment*. As an example I can mention Klinefelter's syndrome. Individuals with this syndrome are males with one or more extra *X* chromosomes, and it seems quite unlikely that any alteration of the environment can prevent the development of this disease. Of course diseases are never monocausal, and the abnormality itself may be determined by a multitude of factors, some environmental and other genetic, but that is a different story. Diseases like Klinefelter's syndrome are genetic diseases in the strongest sense.

However, the definition is not problem-free and I must make one comment. I used the word abnormality, and, as Hesslow points out, "normal" has many meanings. In this case I refer to normality in its statistical sense, e.g., that which is found in, say, 95% of the population (the population being all human beings, all members of the male or female sex or a racial subgroup of humans). If you object that this statistical definition excludes the possibility of a common genetic disease, I may instead defend some teleological view of normality, like a genetic species design [1], but I do not think that the problem is important in this context. Then there are cases where *the genetic abnormality is a necessary but insufficient determinant of the disease and where the environmental factors which complete the causal complex occur normally*. Phenylketonuria may serve as an example, as it is determined by a genetic abnormality and by a normally occurring environmental factor, i.e.,

195

L. Nordenfelt and B.I.B. Lindahl (eds.), Health, Disease, and Causal Explanations in Medicine, 195–197.
© 1984 *by D. Reidel Publishing Company.*

the ingestion of phenyl alanine. I think it is also reasonable to label this a genetic disease (in a weaker sense), as the genetic factor is a necessary determinant, but of course the statement is to some extent tautologous, as we have chosen this genetic abnormality to define a disease entity. Primary lactose malabsorption in a Dane may also serve as an example, as alactasia is rare among Danes and a high consumption of milk products in Denmark is high.

Then we have diseases where the genetic abnormality is a necessary but insufficient determinant and where the environmental factors which complete the causal complex are rare. As an extreme example I can mention suxamethonium sensitivity which is seen in people with pseudocholinesterase deficiency and which may lead to paralysis and ventilation arrest during surgery. This inborn error of metabolism is rare and it seems to have had no harmful effects until the introduction of this muscle relaxant. Suxamethonium sensitivity can be labelled both a genetic and an environmentally induced disease, as both the genetic factor and the environmental factor (suxamethonium treatment) are necessary and abnormal (occur rarely). The fourth possibility is that the genetic factor is an unnecessary and an insufficient determinant (i.e., an inus factor [2]). Anchylosing spondylitis may serve as an example. The genetic factor in this case is HL-A B27 which confers a susceptibility to the development of anchylosing spondylitis which is 127 times that in the general population. Therefore, anchylosing spondylitis is partly genetically determined, but it cannot be called a genetic disease, as the genetic abnormality is not a necessary causal factor. Tuberculosis also belongs to this type of disease, as the genetic determinants are neither necessary nor sufficient, but in this case the disease is defined in such a way that the environmental determinant (the tubercle bacillus) is a necessary causal factor. The status of the disease was just the same forty years ago when it was much more common, and therefore I do not find that it is reasonable to suggest (as hinted by Hesslow) that it was ever a genetic disease.

Finally, there are diseases that are not genetically determined at all, which means that some environmental factor is both a necessary and a sufficient determinant. As an example I can mention arsenic poisoning, as arsenic will cause disease in all human beings. It would of course be quite unreasonable to say that arsenic poisoning is determined genetically, as the causal field must be limited to the genetic constitution seen in human beings. Lactose malabsorption among Eskimos in Greenland has the same status, as alactasia is no abnormality among Eskimos and as the environmental factor (milk consumption) is necessary.

I hope to have shown by these examples that the concept of genetic diseases is a reasonable one. According to the strong definition a disease is genetic, if the genetic abnormality is both a necessary and a sufficient causal factor, which means that the disease will develop in any environment. The minimal requirement is that the genetic abnormality is a necessary determinant, because that means that the genetic factor is a determinant in all patients who are said to have the disease. Other diseases are partly determined genetically, as the genetic abnormality is only an inus factor. In those cases the genetic abnormality is only part of the causal complex in some of those patients who are said to have the disease.

As a last remark I should like to stress that Hesslow's paper, as well as my comment and my own paper at this symposium on the causal basis of the disease classification suffer from the drawback that they are based solely on the Humean or successionist concept of causality. They only discuss the logical interrelationship between causes and effects. Most doctors, however, are philosophical realists and not phenomenalists. They do not confine their interest to the study of successions of events, but wish to understand the disease mechanisms. From that point of view it seems natural to say that deficiency of phenyl alanine hydroxylase is the cause of phenylketonuria, as they understand that this deficiency must lead to an accumulation of phenyl alanine and its metabolites, which again leads to cerebral damage. On the other hand, it is much less acceptable to state that HL-A B27 is one of the causes of anchylosing spondylitis, as we have no knowledge of the mechanism. It is quite possible that there is no direct causal connexion. If we think in terms of the generative theory of causality, we gain the additional advantage that we need not worry so much about the concept of normality. The combination of alactasia and lactose ingestion produces diarrhoea, and the causal mechanism is just the same in Denmark and Greenland.

Herlev University Hospital,
Denmark

BIBLIOGRAPHY

[1] Boorse, C.: 1977, 'Health as a Theoretical Concept', *Philosophy of Science* **44**, 542–73.
[2] Mackie, J. L.: 1965, 'Causes and Conditions', *American Philosophical Quarterly* **2**, 245–64.

C. CAUSAL EXPLANATION

KAZEM SADEGH-ZADEH

A PRAGMATIC CONCEPT OF CAUSAL EXPLANATION*

1. INTRODUCTION

Too much philosophical ink has been spilled on causality since Aristotle. But
the problems remain with us as they were before him. The elimination of
the notion of causality and all of its derivatives from the human language
is probably the only satisfactory solution to these problems. I personally
would prefer this elimination to all philosophizing on causality. The reason
for my preference is this. Knowledge about causes is commonly viewed
as a necessary condition of efficient action, and philosophizing on causality
is justified by saying that knowledge about causes presupposes a concept
of causality. In my opinion, the first part of this view, that is, the supposition
of an action-theoretic demand for causal knowledge, is a dogma; hence there
is no action-theoretic justification of philosophizing on causality. That does
not mean that philosophizing on causality is practically irrelevant. On the
contrary, even to show that philosophical inquiry into the problems – and
pseudoproblems – of causality is a luxury that does not pay would be practically relevant at least in that it would perhaps contribute to a meaningful
redistribution of research grants. Another practical consequence of such an
inquiry would be the reduction of the plentiful 'causal' nonsense produced
in the empirical sciences, especially in medicine. One of my aims in presenting
the following thoughts on causal explanation is to demonstrate how many
of the "causal explanations" offered in medicine belong to this category.

I shall propose a concept of causal explanation which is a pragmatic one
for the following three reasons. It is relative to the three parameters *language,
time*, and the *individual* who renders a causal explanation. To this end, a few
terminological preliminaries are in order.

2. TERMINOLOGY I

I shall first sketch (1) the well-known H–O scheme of scientific explanation
which was proposed by Hempel and Oppenheim in 1948, and (2) Hempel's
concept of causal explanation which is based on this H–O scheme. My
pragmatic concept of causal explanation discussed later is a modified version

201

*L. Nordenfelt and B.I.B. Lindahl (eds.), Health, Disease, and Causal Explanations in
Medicine*, 201–209.
© 1984 *by D. Reidel Publishing Company.*

of Hempel's proposal. Throughout the discussion, I shall refer to a standard object language of the first order, that is, a language in which the quantifiers ∀ ("for all", "for every") and ∃ ("there exists", "thers is") are applied only to individual variables. Our quantified sentences will therefore be of the structure $\forall x \alpha$ or $\exists x \alpha$ where α is any sentence of the language under discussion. Individual variables are represented by $x, y, z, \ldots$; individual constants by $a, b, c, \ldots$; and predicates by $P, Q, R, \ldots$. The Greek letters $\alpha, \beta, \gamma, \ldots$ will be used as statement variables. We read $\neg \alpha$ as "not α", $\alpha \wedge \beta$ as "α and β", $\alpha \rightarrow \beta$ as "if α then β".

A singular statement is a sentence without quantifiers and individual variables. A general, or universal, statement is a sentence of the form $\forall x \alpha$, for example, $\forall x P x$; and a general conditional is a general sentence of the form $\forall x (\alpha \rightarrow \beta)$, for example, $\forall x (P x \rightarrow Q x)$. A general conditional $\forall x (\alpha \rightarrow \beta)$ which is not equivalent to a singular statement is referred to as a "law".

3. H–O SCHEME OF SCIENTIFIC EXPLANATION

According to Carl Gustav Hempel and Paul Oppenheim [3], the scientific explanation of an event provides a particular answer to a why-question of the type "why does the event such-and-such occur?", for instance, 'why does this patient show the symptom S?'. Again, according to their proposal, the answer to such a why-question, that is, the explanation of an explanandum e, takes the form of a logical argument:

$$\left. \begin{array}{l} L_1, \ldots, L_m \\ A_1, \ldots, A_n \end{array} \right\} \text{Explanans}$$
$$\overline{ E }$$

whose conclusion is the statement E which describes the explanandum e. The premises of this argument consist of two types of statements: (1) the laws $L_1, \ldots, L_m$, and (2) certain singular statements $A_1, \ldots, A_n$, called antecedent statements. They describe particular conditions which are realized prior to, or at the same time as, the explanandum. Thus, we have the so-called H–O scheme consisting of the explanans and explanandum statement such that the line indicates the logical relation of deducibility. For example, the simple argument

$$\forall x (D x \rightarrow S x)$$
$$\underline{D a}$$
$$S a$$

represents an H–O explanation of the event Sa which says that the individual a shows the symptom S. This explanandum statement Sa follows from the law $\forall x(Dx \rightarrow Sx)$ expressing that every individual x who suffers from disease D shows the symptom S, and the antecedent statement Da which says that this individual a suffers from disease D. The H–O scheme is also called deductive-nomological explanation, or, DN explanation, because it contains laws and a deductive relation.

The inventors of the H–O scheme have formulated four criteria of adequacy which any scientific explanation should fulfill. Briefly stated, they say: (1) The explanans must contain general laws; (2) The explanans must have empirical content, that is, it must be capable of test by experiment or observation; (3) The explanandum statement must be a logical consequence of the explanans; and (4) The statements constituting the explanans must be true.

I do not want to go into detail here. But it should be emphasized that the last criterion, 'the truth of the explanans', cannot be fulfilled because we can never know whether a general conditional constituting a law is true. This epistemic objection is not the only and not the main one to the H–O scheme. I shall indicate later that there are some serious objections even at the syntactical level and therefore the H–O scheme cannot be accepted in its present form.

4. HEMPEL'S CONCEPT OF CAUSAL EXPLANATION

Hempel distinguished later [2] between *laws of coexistence* and *laws of succession* in the following manner. A law of the form $\forall x(\alpha \rightarrow \beta)$, for example

$$\forall x(Px \rightarrow Qx),$$

is said to be (1) a law of coexistence if the states of affairs described by the antecedent Px and the succedent Qx occur simultaneously, and (2) a law of succession if the antecedent Px describes a state of affairs which precedes the state of affairs described by the succedent Qx. According to Hempel, a deductive-nomological argument of the H–O form

$$\frac{L_1, \ldots, L_m}{A_1, \ldots, A_n}$$
$$E$$

is a causal explanation if the laws $L_1, \ldots, L_m$ are laws of succession. The event described by the total sum of the antecedent statements $A_1, \ldots, A_n$ is viewed as a *cause* of the explanandum described by E.

5. INADEQUACY OF THE H–O SCHEME AND OF HEMPEL'S CONCEPT

The H–O scheme has been widely discussed in philosophical literature. Besides enthusiasm, it has also evoked several earnest criticisms which show why the proposal is not acceptable. I shall not discuss the criticisms in detail. But a few of them should perhaps be briefly summarized in order to see that in spite of its enormous intuitive appeal, the H–O scheme is unsound and therefore unacceptable:[1]

(1) The first and most lethal criticism has been put forward by Eberle *et al.* [1] who have been able to prove logically that the H–O scheme is *trivial* in the following sense. If we accept the H–O scheme, we can explain *every* event by *every* law which is absolutely irrelevant to the event to be explained. For example, it would be possible to explain the fact that a particular patient suffers from a particular disease by the law that 'all iron exposed to oxygen rusts'. The proof is complicated and cannot be demonstrated here.

(2) A law of the form $\forall x(\alpha \to \beta)$ is logically equivalent to the contraposition $\forall x(\neg\beta \to \neg\alpha)$, for example, $\forall x(Px \to Qx)$ to $\forall x(\neg Qx \to \neg Px)$. On this logical basis, Wolfgang Stegmüller ([9], p. 761) has pointed out that if an H–O argument of the form

$$\forall x(Px \to Qx)$$
$$\underline{Pa}$$
$$Qa$$

is accepted as an explanation of Qa, then the H–O argument

$$\forall x(Px \to Qx)$$
$$\underline{\neg Qa}$$
$$\neg Pa$$

should be accepted as an explanation of the negative event $\neg Pa$. But an explanation of this kind is absurd. For instance, if you explain a complaint of a patient with reference to the fact that he is suffering from a particular disease, you cannot reasonably explain the absence of the disease with reference to the absence of the complaint.

(3) The third critic, Wesley Salmon ([7], [8]), says that according to the H–O scheme and Hempel's concept of causal explanation, events may be considered as causes which, however, are causally irrelevant to the explanandum. Consider one of Salmon's examples:

John Jones avoided becoming pregnant during the past year, for he has taken his wife's birth control pills regularly, and every man who regularly takes birth control pills avoids pregnancy ([7], p. 178).

This example is of the following structure:

$$\forall x (Mx \wedge Bx \rightarrow \neg Px)$$
$$\underline{Ma \wedge Ba} \qquad\qquad (*)$$
$$\neg Pa$$

It satisfies all criteria of the H—O scheme, and is thus an excellent H—O explanation but nevertheless absurd. Independent of taking or not taking birth control pills, no male becomes pregnant. We would therefore prefer to explain the fact that John Jones did not become pregnant by the H—O argument

$$\forall x (Mx \rightarrow \neg Px)$$
$$\underline{Ma}$$
$$\neg Pa$$

rather than by the H—O argument (*). The absurd example (*) demonstrates that the H—O scheme violates Occam's razor which I want to reformulate in the following form: Causae non sunt multiplicandae praeter necessitatem ([5], p. 88).

The three criticisms briefly outlined may convince us that the H—O scheme of scientific explanation is inadequate. Consequently, we cannot accept Hempel's concept of causal explanation because it is a mere specialization of the H—O scheme.

6. A PRAGMATIC CONCEPT OF CAUSAL EXPLANATION

The inadequacy of the two proposals discussed above is mainly due to the concepts of 'law' and 'law of succession' employed by Hempel and Oppenheim. I shall therefore propose another concept of 'causal law'. It is constructed in such a manner that none of the objections known hitherto applies to my proposal. A few terminological conventions must be introduced first.

6.1. Terminology II

Let S be a semantical system formulated in a first-order language, that is, an interpreted language of the first order. The individual constants of S denoting time instants may be symbolized by $t, t_1, t_2, \ldots$; and the time variables of S representing any arbitrary t_i may be written $\tau, \tau_1, \tau_2, \ldots$. If P is an n-place predicate of S, the sentences $P(x_1, \ldots, x_{n-1}, \tau)$ and $\neg P(x_1, \ldots, x_{n-1}, \tau)$ are called *state descriptions* in S. If α and β are state descriptions in S, $\alpha \wedge \beta$ is a state description in S. Thus, state descriptions in S are simple statements or conjunctions, and represent simple or complex events occurring at particular instants of time.

If α is a simple state description of the form $P(x_1, \ldots, x_{n-1}, \tau)$ or $\neg P(x_1, \ldots, x_{n-1}, \tau)$, the set $\{\tau\}$ is referred to as the time set of α and written $T(\alpha)$; and the set $\{P\}$ is referred to as the predicate set of α and written $\mathrm{Pred}(\alpha)$. If $\alpha \wedge \beta$ is a state description, $T(\alpha \wedge \beta) = T(\alpha) \cup T(\beta)$, and $\mathrm{Pred}(\alpha \wedge \beta) = \mathrm{Pred}(\alpha) \cup \mathrm{Pred}(\beta)$.

Let SL be the extended semantical system $S \cup L$ where L is any system of predicate logic added to S. If α and β are state descriptions in SL with the free individual variables $x_1, \ldots, x_m, \tau_1, \ldots, \tau_n$, then γ is a *deterministic law of succession* in SL if and only if (1) $\gamma \equiv \forall x_1 \ldots \forall x_m \forall \tau_1 \ldots \forall \tau_n (\alpha \to \beta)$; (2) γ is an empirical sentence; (3) Every $\tau_i \in T(\beta)$ is later than every $\tau_j \in T(\alpha)$; (4) Every predicate $P \in \mathrm{Pred}(\alpha)$ is extensionally different from every predicate $Q \in \mathrm{Pred}(\beta)$.[2]

For the sake of convenience, the quantifier prefix $\forall x_1 \ldots \forall x_m \forall \tau_1 \ldots \forall \tau_n$ of a deterministic law of succession is written Π. If γ is a deterministic law of the form $\Pi(\alpha \to \beta)$, α and β are respectively referred to as the antecedent and succedent of γ, symbolized by $\mathrm{Ante}(\gamma)$ and $\mathrm{Suc}(\gamma)$.

The sentence $\alpha_1 \wedge \ldots \wedge \alpha_n \backslash \alpha_i$ is the conjunction $\alpha_1 \wedge \ldots \wedge \alpha_n$ minus the ith link α_i. If $\Pi_1(\alpha \to \beta)$ is a deterministic law of succession in SL, then α_i is deterministically relevant to β with respect to $\alpha \backslash \alpha_i$ if and only if $\neg \Pi_2 (\alpha \backslash \alpha_i \to \beta)$ is true in SL.[3]

A statement γ of SL is a *causal law* in SL if and only if γ is a deterministic law of succession in SL and every $\alpha_i \in \mathrm{Ante}(\gamma)$ is deterministically relevant to $\mathrm{Suc}(\gamma)$ with respect to $\mathrm{Ante}(\gamma) \backslash \alpha_i$.

6.2. *Causal Explanation in SL*

It is conceivable that a person x offers a 'causal explanation' of an event which you do not accept and vice versa. Since inter-personal differences of this kind are common in science and everyday life, and will not cease to be common in the next millenia, I think it is unreasonable to believe that there are 'true causal explanations' which every person should accept. A pragmatically relativized concept of causal explanation seems to me more reasonable. I am therefore proposing the following concept.

EX is a causal explanans of the event e relative to SL, x, and τ if and only if there are $G_1, \ldots, G_m, A_1, \ldots, A_n$ such that
(1) $EX = \langle G_1, \ldots, G_m, A_1, \ldots, A_n \rangle$;
(2) $G_1, \ldots, G_m$ are causal laws in SL;
(3) $A_1, \ldots, A_n$ are state descriptions in SL;
(4) There is a state description E which describes the event e;
(5) $G_1, \ldots, G_m, A_1, \ldots, A_n$ and E are accepted by x at τ;
(6) Every $\tau_i \in T(E)$ is later than every $\tau_j \in T(A_k)$, where $1 \leqslant k \leqslant n$;
(7) E is SL-deducible from EX;
(8) E is not SL-deducible from a proper subset of EX.

EXP is a causal explanation of the event e relative to SL, x, and τ if and only if there are an EX and a state description E such that
(1) $EXP = \langle EX, E \rangle$;
(2) EX is a causal explanans of the event e relative to SL, x, and τ;
(3) E describes e.

As I mentioned above, this proposal is immune to all the objections put forward to the H–O scheme and Hempel's concept. It has many philosophical consequences which I do not want to discuss here. One apparent consequence, for example, is that a set of statements which is a causal explanation for me may not be a causal explanation for you, and *vice versa*, if you and I subscribe to different systems of knowledge.

On the basis of my proposal, singular and general causes may be distinguished in the following way.

6.3. Singular Causes

If $\langle G_1, \ldots, G_m, A_1, \ldots, A_n \rangle$ is a causal explanans of the event e relative to $\langle SL, x, \tau \rangle$, then the event described by the conjunction $A_1 \wedge \ldots \wedge A_n$ is a singular cause of the event e relative to $\langle SL, x, \tau \rangle$. Obviously, singular causes are relative to conceptual systems, persons and time. It is also clear that for one and the same event e, a set of independent singular causes $\{e_1, \ldots, e_r\}$ may exist. This is the case exactly when there are many different causal explanations of e relative to $\langle SL, x, \tau \rangle$.

6.4. General Causes

Of course, in medicine and philosophy of medicine the term 'cause' is used in another sense. If, for example, Robert Koch says that 'the cause of tuberculosis is the bacillus such-and-such', he means a general cause. In my framework, this term may be defined in the following way. If $\Pi(\alpha \rightarrow \beta)$ is a causal law in *SL*, for instance,

$$\forall x \ \forall \tau_1 \ \forall \tau_2 (Px \ \tau_1 \rightarrow Qx \ \tau_2),$$

the state of affairs described by the antecedent α is a general cause of the event described by the succedent β. Analogous to singular causes, a set of independent general causes of one and the same state of affairs may exist.

7. CAUSAL EXPLANATION IN MEDICINE

In discussing causal explanations in medicine, one should differentiate between searching for general causes and searching for singular causes. A search for general causes is a search for causal laws and is thus a task of medical research. A search for singular causes is a search for causal explanations in dealing with singular cases, that is, with patients in clinical practice. Thus, in the diagnostic process causal explanations may be rendered to obtain diagnoses as antecedent statements $A_1, \ldots, A_n$ of those explanations. Causal diagnostics as causal explanation in clinical practice requires however that causal laws are available which connect diseases and symptoms, complaints, etc. But everyone knows how few causal laws exist in medicine. From this knowledge one can conclude that only few causal explanations are possible in clinical practice.[4]

8. STATISTICAL-CAUSAL ANALYSIS

Due to the lack of causal laws in medicine the physician must use statistical laws. However, it is impossible to render causal explanations on the basis of statistical laws because the explanandum statement does not follow from a statistical explanans. In other words, statistical explanations *do not exist*. It is nonetheless interesting to ask whether some kind of 'causal approach' to statistical relations among events is meaningful. Salmon ([7], [8]) and Stegmüller [10] have demonstrated that this question may be answered affirmatively. The method of *statistical-causal analysis* which they have developed is very complicated and cannot be discussed in a few minutes. I have shown in detail elsewhere that it is applicable to the diagnostic process ([4], [5]).

University of Münster,
Federal Republic of Germany

NOTES

* This paper was presented in my absence by my colleague, Werner Morbach. I would like to thank him for his willingness to represent me at the symposium.

[1] For a comprehensive discussion, see Stegmüller ([9], pp. 708–774).

[2] This is not the appropriate place to philosophize about the notion of 'empirical' used in clause 2 of the definition. It suffices to note that an 'empirical' sentence in SL is a contingent statement in SL. For the definition of the latter term, see ([6], p. 169).

[3] This is a modified and generalized version of Stegmüller's ([10], p. 335) proposal.

[4] Cf. [5].

BIBLIOGRAPHY

[1] Eberle, R., Kaplan, D., and Montague, R.: 1961, 'Hempel and Oppenheim on Explanation', *Philosophy of Science* 28, 418–428.

[2] Hempel, C. G.: 1965, 'Aspects of Scientific Explanation', in *Aspects of Scientific Explanation and Other Essays in the Philosophy of Science*, Free Press, New York, pp. 331–496.

[3] Hempel, C. G. and Oppenheim, P.: 1948, 'Studies in the Logic of Explanation', *Philosophy of Science* 15, 135–175.

[4] Sadegh-zadeh, K.: 1978, 'On the Limits of the Statistical-Causal Analysis as a Diagnostic Procedure', *Theory and Decision* 9, 93–107.

[5] Sadegh-zadeh, K.: 1979, *Problems of Causality in Clinical Practice*, unpublished manuscript, in German. Forthcoming as Volume 2 of *Medizin, Ethik & Philosophie*, Burgverlag, Tecklenburg, Germany.

[6] Sadegh-zadeh, K.: 1982, 'Perception, Illusion, and Hallucination', *Metamedicine* 3, 159–191.

[7] Salmon, W. C.: 1970, 'Statistical Explanation', in R. G. Colodny (ed.), *The Nature and Function of Scientific Theories*, University of Pittsburgh Press, Pittsburgh, pp. 173–231.

[8] Salmon, W. C.: 1971, 'Explanation and Relevance: Comments on Greeno's Theoretical Entities in Statistical Explanation', in C. G. Buck and R. S. Cohen (eds.), *Boston Studies in the Philosophy of Science*, Vol. 8, D. Reidel Publishing Company, Dordrecht, pp. 27–39.

[9] Stegmüller, W.: 1969, *Wissenschaftliche Erklärung und Begründung*, Springer Verlag, Berlin.

[10] Stegmüller, W.: 1973, *Personelle und statistische Wahrscheinlichkeit, Zweiter Halbband, Statistisches Schliessen, Statistische Begründung, Statistische Analyse*, Springer Verlag, Berlin.

INGMAR PÖRN

COMMENTS ON SADEGH-ZADEH'S 'A PRAGMATIC CONCEPT OF CAUSAL EXPLANATION'

There seem to be two main types of views on how to solve the characterization problem for scientific explanations. There are those — we might call them logicists — who proceed from the deductive-nomological model of Hempel and Oppenheim and construe scientific explanation as a two-place deducibility relation between explanandum and explanans, the latter being subdivided into a set of laws and a set of antecedent conditions. In the opposing camp we find pragmatists of various kinds. Some pragmatists are favourably disposed towards the deductive-nomological model but try to supplement it by supplying a pragmatic setting in the form of relativizing factors such as an explainer, an explainee, a context of inquiry, a conceptual system, and so on.

In his paper 'A Pragmatic Concept of Causal Explanation', Professor Sadegh-zadeh advances such a pragmatic view in his solution to the characterization problem for *causal* explanations. He is much influenced by Hempel and Oppenheim and by the use Hempel makes of the deductive-nomological model to characterize causal explanation. Sadegh-zadeh retains deducibility and a nomological component, but he modifies matters in order to avoid the serious objections to which Hempel's account is open. The pragmatic setting is specified in terms of three parameters — language, time, and the individual (explainer) — and the essential modification concerns the notion of a causal law. As far as I can see, Sadegh-zadeh is justified in claiming that his proposal overcomes the difficulties connected with Hempel's account. This is substantial progress, made by Sadegh-zadeh with elegance and economy.

It seems to me that with his model Sadegh-zadeh is well on the way to a truth-independent notion of causal explanation. A causal law is a deterministic law of succession which satisfies a condition of deterministic relevance. In order for deterministic relevance to obtain it is necessary that certain existential sentences be true. The truth of these sentences, however, does not entail the truth of the law of succession, nor, for that matter, need the law of succession be true. According to clause (2) in the definition of a deterministic law of succession, a general conditional, or a sentence which is logically equivalent to a general conditional, is a deterministic law of succession only if it is an empirical sentence. Since no special explanation is given

211

L. Nordenfelt and B.I.B. Lindahl (eds.), Health, Disease, and Causal Explanations in Medicine, 211–212.
© 1984 by D. Reidel Publishing Company.

as to how "empirical" should be understood, we may understand the term in its ordinary sense: a sentence is empirical if and only if sense-experience is relevant to the determination of its truth or falsehood. Accordingly, a deterministic law of succession may be a general conditional which is in fact false. So although certain sentences must be true for a deterministic law of succession to meet the requirement of relevance, the law itself need not be true. It is this fact which I have in mind when I say that Sadegh-zadeh is well on the way to a truth-independent notion of a causal explanation.

This notion, when taken in conjunction with Sadegh-zadeh's definition of a singular cause and his definition of a causal explanans, has an important consequence which must be noted: an event may be a singular cause of another even if the covering laws connecting them are false provided that the explainer accepts them as true. In other words, singular causes need not occur in the world about which the explainer frames his hypotheses, they may be found only in his beliefs. (Strictly analogous considerations apply in the case of general causes.)

I have considerable hesitation over the consequence just noted, and I find it desirable to avoid it. Is it possible to do this by modifying clause (2) of Sadegh-zadeh's definition of a deterministic law of succession while retaining the rest of his account?

University of Helsinki,
Finland

D. OTHER TOPICS ON CAUSALITY

ERIK ALLANDER

HOLISTIC MEDICINE AS A METHOD OF CAUSAL EXPLANATION, TREATMENT, AND PREVENTION IN CLINICAL WORK: OBSTACLE OR OPPORTUNITY FOR DEVELOPMENT?

1. INTRODUCTION

In a recent special article in the *New England Journal of Medicine*, Eric J. Cassell [4] from the Department of Public Health, Cornell University Medical College in New York has written about the nature of suffering and the goals of medicine. Dr Cassell emphasizes in his paper that the relief of suffering and the cure of disease must be seen as twin obligations of a medical profession that is truly dedicated to the care of the sick. He also reminds us that the obligation of physicians to relieve human suffering stretches back into antiquity. He is somewhat disturbed by the comparatively small amount of research devoted to this area, the topic of suffering. He points out that despite a certain overlap, there is a distinct difference between pain and suffering as phenomena.

In his paper Cassell identifies some dimensions in the life of the patients which are relevant to the title of this discussion of holistic medicine. No person exists without others. A patient always has relationships. Persons are often unaware of much that happens within them and why. He also says that everyone has a secret life, and I want to quote here directly:

Sometimes it takes the forms of fantasies and dreams of glory; sometimes it has a real existence known to only a few. Within the secret life are fears, desires, love affairs in the past and present, hopes, and fantasies. Disease may destroy not only the public or the private person but the secret person as well. A secret beloved friend may be lost to a sick person because he or she has no legitimate place by the sick-bed. When that happens, the patient may have lost the part of life that made tolerable an otherwise embittered existence or the loss may be only of a dream, but one that might have come true. Such loss can be a source of great distress and intensely private pain [4].

Cassell also points out that everyone has a perceived future and a transcendent dimension, a spiritual life. This is most directly expressed in religion and in the mystic traditions.

This article, published in one of the most respected medical journals in the world devoted to traditional scientific medicine, emphasizes many elements

L. Nordenfelt and B.I.B. Lindahl (eds.), Health, Disease, and Causal Explanations in Medicine, 215–223.

of holistic medicine by highlighting the role of care in medicine as a logical twin of technical cure and treatment.

Here I will examine causality within holistic medicine in general and in its clinical applications and go on to develop some thoughts along those lines. This is an attempt to draw attention to some principles of reasoning that have emerged from broad, but non-systematic reading of various kinds of published material from stenciled pamphlets in vegetarian food shops to periodic popular journals, and books or specific chapters of books. My impression therefore comes from a broad scope of sources, generally not identified as "scientific" in a traditional sense. A thesis from my department [11] published in Swedish has dealt with some of the aspects above. With the limited amount of time external circumstances have allowed me, it has not been possible to systematize these rather heterogenous sources. I am naturally aware of the loss of scientific weight this implies but ask the reader to accept the paper as a preliminary mapping of an area with an extreme diversity of expressions but where some main streams can be identified. With these limitations, I will therefore, by way of introduction, examine some of the underlying concepts in this respect.

As I have intentionally refrained, with a few exceptions, from giving detailed references, my discussion should be looked upon as a biased introduction to a critical study of basic concepts of holistic medicine. The aim of this presentation is therefore to take a first step in the discussion of holistic medicine. Critical specific comments, examples, quotations and references could be a natural second step in the exchange of views. Though internationally based, national variants of holistic medicine are common-place.

2. EXPANSION OF HOLISTIC MEDICINE

There is presently a wave of holistic medicine sweeping over not only the medical establishment as such [7], but also over services only slightly connected with medicine.

The holistic approach is found not only in relation to the medical world but also in a variety of fields, social welfare, education, working life and so on. The holistic view could be perceived as an attitude and also a method to widen the perspective on the individual situation when in crises or disease as well as a method of causal explanation. It is difficult to say exactly what are the concepts of the holistic view, especially of holistic medicine. However, it should obviously be seen as a complement to more technically oriented conventional ways of treatment and prevention.

It has also been claimed that holistic medicine represents a renaissance of more humanistic values in care emphasized in Cassell's article, but these humanistic values have long existed in the care of patients. It is also suggested and actively proposed that holistic medicine means something more than bringing together technological achievement with humanistic values to improve the quality of life.

The holistic principles are not only applied to patient care in ordinary everyday practice, but also to preventive measures taken both for individuals and population groups. A fertile soil for holistic medicine is found in the perceived and real failures, mistakes and incompletenesses of traditional medicine and scientific work ([6], [10]). Our inability to understand and solve present, important problems would support the need for some cross-scientific research. 'Multidimensional eco-philosophy', 'ethics', and 'ecology' are some of the many terms found in the present day discussion on holistic medicine. Other terms are 'biofeedback' and 'behavioural medicine', and several books have been published on the topic. An example on ideas behind holistic medicine is well worth penetrating.

3. THE RULES OF A HOLISTIC MEDICAL ASSOCIATION

The American Holistic Medical Association[1] has defined health as "a state of well-being in which an individual's body, mind, emotions, and spirit are in tune with a natural cosmic and social environment" [1]. Here we are faced with a definition basically of the same concept of health as the one we are familiar with from WHO, "Health is a state of complete physical, mental, and social well-being, and not merely the absence of disease and infirmity" ([13], p. 459).

Among the 21 entries, also divided into sub-entries [1] it is stated in the declaration of the American Holistic Medical Association that a doctor practising holistic medicine should provide "a patient referral service for the public to holistic physicians". There is very little, though, about the actual practice of holistic medicine in relation to patients; instead more of the fight for professional freedom, for instance "to protect physician control over the practice of medicine against increasingly forceful incursions by politicians, bureaucrats, clerks, and administrators" and also "to defend freedom of choice of both physician and types of treatment". Only in point 19 is a clear therapeutic part or a part devoted to prevention to be found, namely " to provide guidelines on nutrition, exercise, stress management, holistic principles for the general public" but also "to work toward a clean environment, free of nuclear waste, toxic metals. . ." etc. An interesting point is point No. 20, where it is stated that holistic medicine should "support the acceptance of spiritual healing and of intuitive diagnosis as scientifically valid". Here it is forcefully maintained that holistic medicine *should* be accepted as a full and equal partner in the scientific medical community. A further recommendation in the same directions is "to work towards a balance between the purely 'scientific' and technological aspects of medical practice with more humanistic and spiritual values". Here, in essence, the rules clearly recommend a balance between classical traditional scientific and technological aspects with humanistic and spiritual values. There are also, finally, suggested lines of research, namely "to support scientific research in all aspects of holistic medicine".

In holistic health models of reasoning, prevention and treatment are included, beyond ordinary medicine, a variety of other fields of health practice like acupuncture, sometimes on the outside of classical medicine and anthroposophy. In contrast to conventional medicine more outspoken aspects of metaphysics are included besides the widely accepted, down-to-earth, everyday advice on physical exercise, a balanced diet, discouragement of smoking and so on. In holistic medicine one often refers to a whole collection of metaphysical systems. There seems to be an inclination to construct metaphysical ad hoc theories.

Terms like lifestyle, psycho-biology, consciousness and various psychological terms are included. The scientific image of holistic medicine is propagated by its journals: the *Journal of Humanistic Psychology*, the *Journal of Transpersonal Psychology* and the *Journal of Behavioural Medicine*.

4. SOME FEATURES OF HOLISTIC MEDICINE: CRITICAL COMMENTS

It is generally claimed by the supporters of holistic medicine that medicine practiced according to these principles is superior to that of grey, everyday medicine with its failures and few brilliant results. Several principal weaknesses could be found in the reasoning on causal relationships of those advocating holistic medicine.

First, they mix the best part of holistic medicine — the well-being orientation — with the best part of the clinical classical medicine — the safe diagnosis and treatment, but leave out the unclarities and negative or difficult or unsolved parts.

Second, they apply the same system to measure results both on individual and population level, e.g., mortality risk of a group and the unpredictability of the course of a disease for a single individual.

Third, there is an unmistakable aspiration or at least wish of holistic physicians to claim that patients have been cured, lived longer, have experienced an improved well-being, have come to work earlier, that their cancers have disappeared. But they claim their cure, the effectiveness of their method in *conventional* terms without using the conventional methods correctly.

Fourth, they apply here the methods of classical scientific medicine in the evaluation of the effect of medical care — increased survival rates, reduced disability, pain and so on. Here they face a usually well-defined collection of difficult problems familiar to clinical and epidemiological research workers.

Fifth, the concept of causality in holistic medicine seems to be based on the permanent mixture of effects of treatment in general with effects in management of individual patients. The general perception of causal principles is mostly without discussion applied to the individual patient. This sort of general reasoning is often followed without having actual knowledge of the factors that rule the life of a patient, including his genetic make-up. This can be exemplified by Laetrile treatment [5], or THX-Treatment in Sweden ([3], [8], [9]), where both "drugs" are popular but have no proven effect. Another example is Iscador [2].

Sixth, the mixture of the care and the cure part of medical treatment is not so unique a feature of holistic medicine as many of its advocators think. The care part and most of the elements that I mentioned in the introduction, namely compassion, help, support, reassurance are definite and well-defined complements to the technical part of the medical work of every physician.

Seventh, there is very little said about which kind of patients should be selected for treatment according to the principles of holistic medicine. Here we have two lines of argument, one that holistic medicine is a general method which should be applied to any patient even those who have cancer, minor illnesses, fractures, rheumatoid arthritis or, that holistic medicine is the method of choice for less well defined diseases like stress-induced conditions, psychological or psychiatric disturbances.

Eighth, privacy is severely intruded upon in holistic medicine. It takes up theoretically every aspect, or at least a large number of aspects of a person's life — fantasies and dreams, loves, hates and failures. It leads to an exposition of private life very similar to that which we find in psychoanalysis. This sacrifice of privacy would only be reasonable if there were a causal relationship between this loss and a successful treatment. Probably people in general will not be willing to accept this part of the principles of holistic medicine. They would, most likely, prefer a technically skillful physician, than to have their private life intruded upon.

Ninth, an objection is that the causal aspects behind a patient's condition will be so complicated, amorphous and complex that it is impossible both for the physician and the patient to understand fully the causal relationships or explanations that lie behind a certain treatment or the perception of changes in the patient's condition. Because of its broad scope it is inevitable that a substantial element of chance and subjectivity is introduced in the interpretation of individual causal relationships. The importance of chance and subjectivity — aspects of reliability — is suppressed or neglected in holistic medicine, when it applies — or tries to apply — traditional scientific methods.

Tenth, my final objection is that holistic medicine always includes definite parts of metaphysical reasoning and/or of mysticism. The causal relationships found here are by definition not explicable within a conventional scientific framework. It can, however, serve as an escape when treatment fails. The basic dilemma is the ambiguity between physical and metaphysical explanations of relationships. This dilemma that faces holistic medicine and the causal principles behind it cannot be tested, either with the methods of holistic medicine itself, especially if there is a clear or implicit suppression of a traditional scientific method, or with ordinary scientific methods because of the clear metaphysical part.

Some examples do, however, exist. Some years ago, in a scientific journal, a controlled study was published on the effect of prayers [12]. A priest read prayers for a random sample from a group of patients. There were no differences in therapeutic success between those exposed and those not exposed to prayers.

Finally, some positive comments. In spite of the many objections I find a fair amount of consensus between the objectives of the art of conventional clinical medicine, and those of holistic medicine. Both should increase well-being, provide confidence and reassurance and create a positive image of the future, as in the article I quoted first. Conventional as well as holistic medicine should consciously or unconsciously provide patients with an image of a future which is *better* that that found according to scientific results. Examples are abundant. To provide the incurable and the dying with hope, those in pain with comfort, and so on. Some suppression of reality is probably necessary for the physician and the patients.

Another aspect of holistic medicine is the use of music, poetry, and art in treatment for, e.g., psychiatric diseases. Here a therapy or experience is introduced that everyone is exposed to in normal life, namely the feeling of joy, satisfaction, and well-being by listening to music or poetry, looking at paintings or going to the theatre. Those factors certainly contribute to well-being and quality of life but within a medical framework such therapy is at best regarded as highly unspecific.

The richness and variety of positive associations provided by holistic medicine makes it less prone to critical evaluation. Holistic medicine will not become a science in its own right just by piling other science or pseudo-sciences in a heap and pretending or stating that they are all in combination part of a successful treatment. The customary unbiased background against which a researcher can look for causal relationships is therefore not present.

Finally, it should not be forgotten that holistic medicine can be *partly* evaluated within traditional scientific research designs. It is also *theoretically* possible to evaluate some of the causal relationships put forward by holistic medicine. Whether this should be done by those practicing holistic medicine or others is an open question, as is also whether it is really a research priority.

5. SUMMARY AND CONCLUSIONS

My critical conclusions on principles of causality in holistic medicine can be summarized under three main headings (to some extent these are interwoven):

A. Aspects of Research in General

A1. In holistic medicine hypotheses and theories are confused with proven results.

A2. Explicitly unclear models are used in evaluation of holistic treatment performed by those that practice holistic medicine.

A3. The cross-scientific approach often used by holistic medicine is valid only to a limited extent, especially in the explanation of causal relationships.

A4. An over-simplification of complex interactions mixed with ignorance or omission of scientifically established results.

A5. The scientific ideal of holistic medicine is very similar to that of a renaissance scientist/scholar, covering all essential knowledge and experience of his time without realizing the colossal volume of present knowledge, and disregarding the historic development.

A6. Holistic medicine claims that holistic treatment also gives an improvement in traditional measures such as prolongation of life and cure of diseases but this is often based on research designs not accepted by scientists working conventionally.

A7. Unmeasurable metaphysical parts with various designations of holistic medicine are systematically given more weight than traditional measurable quantities.

A8. The metaphysical part of holistic medicine implies a partial rejection of ordinary principles of causality.

B. *The Application of Scientific Methods in the Evaluation of Treatment*

B1. Holistic medicine extrapolates from unproven, highly positive value-loaded theory to actual treatment of individual patients.

B2. In its general approach holistic medicine has more in common with the doctrine of salvation than conventional medical practice and cannot therefore be tested according to conventional scientific principles.

B3. The ultra-individual treatment provided idealistically to individual patients in the practicing of holistic medicine logically cancels or reduces the possibility of interpreting a causal relationship between treatment and result.

C. *The Clinical Practice of Holistic Medicine*

C1. Holistic medicine confuses a variety of research objectives with those of individual treatment.

C2. Implicitly holistic medicine is not research- but practice-oriented. That is, the causal relationships that holistic medicine claims are accepted as factual and the main issue is practice and spreading the gospel.

C3. Holistic medicine can sometimes mean a rigid application of a standard approach to conditions disregarding or suppressing well-established causal relationships and the natural history of disease.

C4. In holistic medicine the values of care found in all conventional medical practice (to comfort, to entertain hope, to reassure, to support) are mixed with the technical part (prescribing drugs, measuring blood-pressure, performing diagnostic X-ray, operating, etc.). Basically holistic medicine means "more of the same" and in principle little new in treatment.

C5. The principles of causality within holistic medicine implies an intrusion of the patient's privacy, far more extensive than that of conventional medicine, in order to attain a so-called better result.

C6. Holistic medicine wants to stay within the borders of conventional medicine to be able to attain its benefits (e.g., eligible for health insurance benefits, cooperation with and within traditional medicine).

Huddinge University Hospital,
Sweden

NOTE

1 Dr Björn Liedén, Mönsterås, Sweden, kindly provided me with a statement of goals, purposes, and planned membership services of the American Holistic Medical Association, and some other sources for which I am most grateful.

BIBLIOGRAPHY

[1] American Holistic Medical Association: *Goals, Purposes and Planned Membership Services of the American Holistic Medical Association*, 6932 Little River Turn-Pyke, Annadale, Virginia.

[2] Anonymous: 1981, 'Antroposofer hos hälsovårdsministern om Iscador' ('Anthroposophers Visit the Minister of Health About Iscador', *Läkartidningen* 78, 3000.

[3] Anonymous: 1981, 'THX-klinikerna öppnade på nytt efter "ansökan" om registrering' ('THX-Clinics Opened Again After "Application" About Registration'), *Läkartidningen* 78, 2811.

[4] Cassell, E. J.: 1982, 'The Nature of Suffering and the Goals of Medicine', *The New England Journal of Medicine* 306, 639–645.

[5] Classileth, B. R.: 1982, 'After Laetrile, What?', *The New England Journal of Medicine* 306, 1482–1484.

[6] Cochrane, A. L.: 1973, *Effectiveness and Efficiency – Random Reflections on Health Services*, The Rock Carling Fellowship, 1971 – The Nuffield Provincial Hospital Trust, 3rd.

[7] Editorial: 1979, 'Mind and Cancer', *The Lancet* 1, 706–707.

[8] Editorial: 1980, 'Lex THX – för vem?' ('Lex THX – For Whom?'), *Läkartidningen* 77, 2751–2756.

[9] Einhorn, J.: 1980, 'Propositionen om naturmedel för injektion, restriktivt förslag som kan sanera marknaden' ('The Proposition About Naturopathic Medicine for Injection, Restrictive Proposal that Can Clear the Market'), *Läkartidningen* 77, 4236–4238.

[10] Illich, I.: 1975, *Medical Nemesis*, Calder and Boyars Ltd., London, U.K.

[11] Jacobson, N.-O.: 1979, 'Naturläkemedel och okonventionella behandlingsmetoder' ('Naturopathic Medicines and Unconventional Methods of Treatment'), Thesis, Department of Social Medicine, Huddinge University Hospital, Karolinska Institute, Stockholm.

[12] Nelson, B.: 1965, 'The Objective Efficacy of Prayer', *The Journal of Chronic Diseases* 18, 367–377.

[13] World Health Organization: 1958, *The First Ten Years of the World Health Organization*, Geneva, Switzerland.

STUART F. SPICKER AND H. TRISTRAM ENGELHARDT, JR.

CAUSES, EFFECTS, AND SIDE EFFECTS: CHOOSING BETWEEN THE BETTER AND THE BEST

> "If we had no drugs, the public would not suffer any adverse drug effects . . . ." ([17], p. 23.) – *Louis Lasagna, M.D.*

1. CAUSING, FORESEEING, AND INTENDING

Throughout the history of Western thought, philosophers and physicians have explored the nature of causation and the general relation that obtains between effects and causes. They have sought to discover causal relationships and even causal laws. Physicians have, however, not conducted themselves as pure scientists. They have, instead, attended to causes through a grid of value judgments embedded in particular practices of curing and caring. As has been argued in comparisons of law and medicine, these non-epistemic goals direct the language of causation in applied disciplines [4].

In practically directed activities, we select particular causes for attention out of a background of causal conditions ([4], p. 125). Thus, a public health physician may speak of poor economic conditions as the cause of widespread tuberculosis whereas a clinician is likely to speak of *mycobacterium tuberculosis* as the cause of tuberculosis, although it is only a necessary not a sufficient condition ([4], p. 123). The selection of the causal conditions for attention turns not only on epistemic interests, but on the practical concerns of an applied science. It then not infrequently happens that a variety of particular causes are selected for a variety of practical, useful, and even merely prudential reasons. "Causal language" is often employed to serve particular social ends ([4], p. 127). Medicine's interest in causes, as this volume shows, is tied to a number of practices directed to achieving both special non-moral goods and avoiding special non-moral evils: preserving health and restoring functions, avoiding pain and deformity.

However, the character of human anatomy and physiology is such that the achievement of these goods is usually at a price. In curing a patient of cancer, a breast or leg is often sacrificed. In treating carcinoma of the prostate, impotence may be a "side effect", or in treating an infection, there may be an allergy as a "side effect" of the antibiotic therapy. A field of causal conse-

225

L. Nordenfelt and B.I.B. Lindahl (eds.), Health, Disease, and Causal Explanations in Medicine, 225–233.
© 1984 by D. Reidel Publishing Company.

quences is thus interpreted in terms of primary goals and significant, but often tolerable, collateral effects. In the cases noticed above, the cure of cancer and of the infection are as truly *effects* of the interventions as are disfigurement, lameness, impotence, or the allergic reaction. However, the field of effects has been organized in terms of a schema of human interests that interprets the significance of these effects. It is this schema that we intend to address in this essay. The practice of denominating some effects as *"side"* effects in contrast with what one might term *primary* effects is an epiphany of the pragmatic character of causal language in medicine.[1] This practice marks some of the effects of medicine's interventions as foreseen and intended, others as foreseen, regretted but tolerable — yet others as foreseen, regretted and to be blunted by further interventions, or not to be tolerated. Such interpretations of effects presuppose an economy of costs and benefits which characterizes the significance of causal outcomes in medicine.

The pharmacologic revolution of the last quarter century not only has had a positive effect on patients, but it has also been a source of injury for many patients. From the rather direct effects of iatrogenically induced disease and illness to patient-induced drug overdose or noncompliance (topics which shall not be addressed here), the pharmacologic revolution and its litany of therapeutic regimens which are so defining of Western medicine have generated a plethora of untoward effects, adverse drug reactions. Though these have become more noticeable, they have a history as old as medicine itself. One of the earliest records of such drug-related responses can be found in Hippocrates' description of the side effects of oxymel[2] in *Regimen in Acute Disease*:

You will find the drink called oxymel often useful in acute diseases, as it brings up sputum and eases respiration. The occasions, however, for it are the following. When very acid it has no slight effect on sputum that will not easily come up; for if it will bring up the sputa that causes hawking, promote lubrication, and so to speak sweep out the windpipe, it will cause some relief to the lungs by softening them. If it succeeds in effecting these things it will prove very beneficial. But occasionally the very acid does not succeed in bringing up the sputum but merely makes it viscid, so causing harm. It is most likely to produce this result in those who are mortally stricken, and have not the strength to cough and bring up the sputa that block the passages. So with an eye to this take into consideration the patient's strength, and give acid oxymel only if there be hope. If you do give it, give it tepid and in small doses, never much at one time.

But slightly acid oxymel moistens the mouth and throat, brings up sputum and quenches thirst. It is soothing to the hypochondrium and to the bowels in that region. It counteracts the ill effects of honey, by checking its bilious character. It also breaks flatulence and encourages the passing of urine. In the lower part of the intestines, however, it tends to produce moisture in excess and discharges like shavings. Occasionally in acute diseases this character does mischief, especially because it prevents flatulence form passing along, forcing it to go back. It has other weakening effects as well, and chills the extremities. This is the only ill effect worth writing about that I know can be produced by this oxymel ([9], pp. 113, 115).

2. A SCHEME FOR UNTOWARD CLINICAL EVENTS

Untoward clinical events have a diverse character: (1) accidental poisoning, (2) suicides "caused" by drugs, (3) victims of drug abuse who intentionally or accidentally does or overdose themselves ([23], p. 10), (4) therapeutic failures, persons who fail to take the prescribed and recommended drug regimen, and (5) patients who become ill due to the recommendations by physicians of various drug regimens (iatrogenically induced) or through surgical or other therapeutic interventions ([12], p. 24).[3] All of these identify therapeutic outcomes which because of various values are taken as events to be avoided. Here in particular we shall examine the case of adverse drug reactions as a special example of untoward clinical events in order to illustrate how therapeutic interventions (causes) are tied to effects which are then interpreted in terms of practices which assign special significance to such outcomes ([7], [10], [18], [20]). These practices presume a nexus of judgments regarding (1) responsibility for foreseen, costly effects that are collateral with therapeutic goals, (2) the ranking of untoward side effects in terms of their costliness in function, pain, and financial expense, and (3) the role the patient should play in establishing such rankings.

A review of the myriad of events which fall into this category of untoward clinical events reveals a broad range of outcomes including serious injury as well as mild "side effects", the ones that are, as one writer put it, a "nuisance" ([23], p. 11). It should be noted that both the serious as well as the mild side effects can be the result of known *and* unknown actions of a drug. Adverse drug reactions include at one extreme those that are (i) fatal and (ii) severe and non-fatal and at the other extreme side effects that are (iii) moderate and (iv) mild,[4] "innocuous", or "negligible" ([6], p. 110). Among such "effects" are noted the following: weakness, belching, blurred vision, daytime drowsiness, palpitation, malaise, anorexia, jaundice, itching, nausea, vomiting, abdominal distress, dry mouth, diarrhea, skin rashes, fever, headache, dizziness, syncope, dark urine, urinary frequency, sweating, constipation, eosinophilia, hyperacidity, thrombophlebitis, peripheral neuropathy (numbness and paresthesia of an extremity) as well as cancer, dreams, and impotence ([8], p. 257; [12], p. 21; [13], p. 1240). To view these outcomes as side effects presupposes a set of goals in terms of which one can distinguish such from primary effects. The concept of side effects presupposes a particular value grid. This is seen when one explores observations such as that of Robert Veatch, which might on first examination appear paradoxical. "It is reported that the Spanish Pharmacopia describes estrogen-progesterone combinations as effective in regulating menstrual cycles, but as having a serious side effect of preventing pregnancy" ([22], p. 76). Side effects stand out against a background of expected outcomes and accepted evaluations.[5]

In 1974 a representative of the Pharmaceutical Manufacturers Association, in response to somewhat inadequate data, calculated that a reasonable esti- mate of the number of deaths due to adverse drug reactions in the United States in patients suffering form non-lethal diseases was probably between two and three thousand. In addition, some 160 000 patients were admitted to hospitals yearly for drug-induced illness in a context of some 32 million patient admissions ([20], p. 1043). In short, the choice of therapies and the decision as to whether particular therapeutic agents should even be employed requires a series of serious, difficult, and often precarious calculations weighing possible benefits and possible harms. They presuppose complex, often under- examined value judgments. Consider, for example, the reasoning of Dr Ed Cadman regarding the difficulties faced in cancer therapy. With candidness, he remarks: "There is great concern that the effective cytotoxic therapy that allows long clinical remissions of some cancers may induce another. There is little doubt that acute nonlymphocytic leukemia following the therapy of Hodgkin's disease is occurring at an alarming frequency — 109 cases reported by 1976 .... Similar observations have been noted with multiple myeloma .... In addition, the use of alkylating agents has also been associated with the development of acute nonlymphocytic leukemia in chronic lymphocytic leukemia ..." ([2], p. 96). The author continues: "This complication [second malignancies] is unfortunate. But when one contrasts the potential benefits to many patients, especially those who remain free of cancer following the use of cytotoxic drugs, with the few malignancies that may develop, it becomes clear that this is a risk worth taking. The second malignancy is apparently the ultimate price required for success. Because the second malig- nancy is a real threat following long-term chemotherapy, especially the alkylating drugs, one must seriously question their prolonged use in non- malignant diseases that have a projected long survival" ([2], p. 97). Physicians tend to justify the diagnostic or therapeutic use of a drug by determining the favorable ratio of anticipated benefits to potential risks ([15], p. 75). Thus a therapeutic ratio is calculated as the ratio of beneficial to adverse effects ([8], p. 1362).[6] What is especially problematic is that some drugs must be given in doses that approach or even reach the toxic range for some patients ([1], p. 53).

As a result, the existence of adverse side effects reminds us to exercise great care with the therapies that cause them. Justifications for their use must be cast in terms of the ratio of benefits to harms which warrant a drug in a particular dosage. Thus the Hippocratic injunction *primum non nocere* is, as Louis Lasagna remarks, "less apt than it was centuries ago, when there were few potent drugs available" ([17], p. 468). It is now perhaps best rephrased as *benefit more than harm*. Effects must be carefully placed and understood within schemata of judgments concerning worthwhile benefits and costs, quality of consent, and accountability of physicians who are charged with being careful initiators of causal chains. As the late Dr Franz Ingelfinger, whose physician advocacy was never a secret, maintained (in an editorial in the *New England Journal of Medicine*): "Physicians — and the public as well — should not be concerned about all adverse reactions, but rather about the number of unwarranted adverse reactions" ([10], p. 1003). By "unwarranted" Dr Ingelfinger meant only those drug administrations in which the physician took too great a *known* risk with the patient. He was thus indicating a background, often underexamined, view of proper exchanges of costs for benefits. Exploring "causes" and "effects" in medicine leads thus not just to epistemological issues, but to central issues in value theory and bioethics: the ranking of medical harms and goods; the role of patients in establishing such rankings; and the responsibilities of physicians to be careful initiators of causal chains.

3. SIDE EFFECTS AND PRIMARY EFFECTS: A JUNCTION OF EPISTEMOLOGY AND BIOETHICS

As the previous sections show, a study of causes and effects in medicine is at once bound to traditional issues in bioethics. Causes and effects do not exist value-neutrally for an applied science such as medicine. Instead, causes (i.e., therapeutic interventions) must be judged in terms of the moral and non-moral values we assign to their effects (i.e., therapeutic outcomes). In part, these causes and effects are open to straightforward cost/benefit calculations. However, the general difficulties of such calculations show themselves here. Any assessment of costs and benefits presupposes an authoritative schedule of costs and benefits. One must be able to rank costs and benefits in order to judge the worthwhileness of an intervention. However, an absolute schedule does not appear to be available, especially in secular, pluralist societies such as ours. Such a ranking requires an authoritative sense of values. This however brings one to the problem of selecting the proper sense to establish the authoritative ranking of costs and benefits. This will, however, require a further sense, and so on *ad indefinitum*.

As a result, in a peaceable moral community, a common schedule of costs and benefits will need to be negotiated. One will need to fashion an ordering of goods and harms so as to judge commonly the benefits and detriments of particular undertakings. Hence the central role of free and informed consent in medical decision-making in secular, pluralist societies. Free and informed consent in part functions as a means of fashioning an ordering of goods and harms agreed to in order to enable the joint endeavors of a patient and a physician. The goals of therapy which are used to discriminate between *primary* effects (those in accord with the therapist's and patient's goals) and *side* effects (those effects foreign to those goals) are as much created as discovered by physicians and patients.

Epistemology and bioethics thus intertwine. To know what the side effects and the primary effects of therapy are, and to decide when the possible deteriments of the side effects outweigh the possible benefits of the primary effects, one must know the value system of the physician and the patient. Indeed, the notions of 'primary effects' and 'side effects' presuppose a major social practice, which is at the core of much of the debate in bioethics: the physician-patient relationship as a moral undertaking. In medicine, knowing and valuing are conjoined.

This essay, in short, supports the conclusions of the majority of essays in this volume. An understanding of the concepts of causality, health, and disease in medicine cannot be developed on the basis of epistemic considerations alone. Concepts of health and disease in an applied science such as medicine serve to direct action. One must, as a result, frame such concepts not only so that they may be true to reality, but useful as well. Definitions and classifications in medicine tend as a result not to be natural, but artificial, framed with regard to instrumental concerns. Our analysis of the notions of primary effects and side effects sustains this viewpoint. The clinical sciences contrast with the basic sciences, in so far as the latter are seen as pursuits of knowledge for knowledge's sake. The clinical sciences focus on a wide range of human non-epistemic goals and purposes: freedom from pain, the realization of norms of body form and grace, increasing life expectancy, etc. Causes and effects are then highlighted in terms of their usefulness in achieving these goals. Some effects are regarded as side effects, and some side effects as therapeutically unjustified side effects by appeals to these goals. This practice of medicine reflects a rich interweaving of explanations and evaluations.

University of Connecticut   *The Baylor College of Medicine,*
School of Medicine,   *U.S.A.*
U.S.A.

NOTES

[1] The term 'side effects' is employed in a wide range of contexts. Jonathan Glover, for example, in his discussions of euthanasia and abortion, views the impact of the patient's death as having side effects — the "effects upon people other than the one killed" ([7], pp. 40, 113). He adds, "there are the effects on the family and friends of the persons killed" ([7], pp. 114; 186–188). He also remarks that side effects do not necessarily carry less weight than the direct objectives, e.g., killing X. And, of course, side effects can also be beneficial, though the term often misleads and suggests that this is not the case ([7], p. 115).

[2] This concoction mentioned by Hippocrates was apparently a form of vinegar of honey. See *Stedman's Medical Dictionary*: "Vinegar of honey; a mixture of 15 percent each of acetic acid and distilled water in purified honey. Used as a gargle in sore throat and as an excipient" ([21], p. 1157).

[3] We intentionally omit here any discussion of a physician's reluctance to admit the limits on his/her ability to modify many conditions or to properly set a therapeutic end point for the drugs he/she prescribes. See ([18], p. 136).

[4] See ([6], p. 113; [11], p. 576 and [14], p. 248).

[5] Also see [11], p. 575; [13], p. 1236; [19], p. 240; [20], p. 51; and [23], p. 10. An ADR is also defined as an undesirable clinical manifestation . . . that is consequent to and caused by the administration of a particular drug ([16], p. 623).

[6] ([18], p. 1364.) Also see ([6], pp. 114, 119), ([12], p. 15): and ([15], p. 78), ([20], p. 1044).

BIBLIOGRAPHY

[1]   Anonymous: 1980, *American Medical Association Drug Evaluation*, 4th ed., John Wiley and Sons, New York, pp. 18–53.

[2]   Becker, F. F. (ed.): 1977, *Cancer — A Comprehensive Treatise*, V, Plenum Press, New York, pp. 97, 638 *et passim.*

[3]   Edelstein, L.: 1967, *Ancient Medicine: Selected Papers of Ludwig Edelstein*, O. Temkin and C. L. Temkin (eds.), Johns Hopkins Press, Baltimore, Maryland.

[4]   Engelhardt, H. T., Jr.: 1981, 'Relevant Causes: Their Designation in Medicine and Law', in S. F. Spicker, J. M. Healey, and H. Tristram Engelhardt, Jr. (eds.), *The Law-Medicine Relation: A Philosophical Exploration*, D. Reidel Publishing Co., Dordrecht, Holland, pp. 123–127.

[5]   Feinstein, A. R.: 1971, 'How do We Measure "Safety and Efficacy"?', *Clinical Pharmacology and Therapeutics* **12** (3), 544–558.

[6]   Feinstein, A. R.: 1974, 'Clinical Biostatistics: The Biostatistical Problems of Pharmaceutical Surveillance', *Clinical Pharmacology and Therapeutics* **16**, 110–123.

[7]   Glover, J.: 1979, *Causing Death and Saving Lives*, Penguin Books, Ltd., Middlesex, England.

[8]   Green, D. M.: 1964, 'Pre-Existing Conditions, Placebo Reactions, and "Side Effects" ', *Annals of Internal Medicine* **60**, 255–265.

[9]   Hippocrates: 1962, *Regimen in Acute Diseases*, in *Hippocrates*, Loeb Classical Library, trans. W. H. S. Jones, William Heinemann Ltd., London, Vol. II, Sections LVIII–LIX.

[10] Ingelfinger, F. J.: 1976, 'Counting Adverse Drug Reactions That Count', *The New England Journal of Medicine* **294** (18), 1003–1004.

[11] Irey, N. S.: 1976, 'Adverse Drug Reactions and Death', *Journal of the American Medical Association* **236** (6), 575–578.

[12] Karch, F. E. and L. Lasagna: 1974, *Adverse Drug Reactions in the United States: An Analysis of the Scope of the Problem and Recommendations for Future Approaches*, A Report prepared for Medicine and the Public Interest, Washington, D.C.

[13] Karch, F. E. and L. Lasagna: 1975, 'Adverse Drug Reactions', *Journal of the American Medical Association* **234** (12), 1236–1240.

[14] Karch, F. E. and L. Lasagna: 1977, 'Toward the Operational Identification of Adverse Drug Reactions', *Clinical Pharmacology and Therapeutics* **21** (3), 247–254.

[15] Koch-Weser, J., E. M. Sellers and R. Zacest: 1977, 'The Ambiguity of Adverse Drug Reactions', *European Journal of Clinical Pharmacology* **11**, pp. 75–78.

[16] Kramer, M. S. *et al.*: 1979, 'An Algorithm for the Operational Assessment of Adverse Drug Reactions: Background, Description, and Instructions for Use', *Journal of the American Medical Association* **242** (7), pp. 623–632.

[17] Lasagna, L.: 1964, 'The Diseases Drugs Cause', *Perspectives in Biology and Medicine* **9**, 457–470.

[18] Melmon, K. L.: 1971, 'Preventable Drug Reactions – Causes and Cures', *The New England Journal of Medicine* **284** (24), 1361–1368.

[19] Naranjo, C. A. *et al.*: 1981, 'A Method for Estimating the Probability of Adverse Drug Reactions', *Clinical Pharmacology and Therapeutics* **30** (2), 239–246.

[20] Stetler, C. J.: 1974, 'Drug-Induced Illness: Letters to the Editor', *Journal of the American Medical Association* **229** (8), 1043–1044.

[21] *Stedman's Medical Dictionary*, 21st ed., Williams and Wilkins Co., Baltimore, Maryland, 1966.

[22] Veatch, R.: 1981, 'Federal Regulation of Medicine and Biomedical Research: Power, Authority, and Legitimacy', in S. F. Spicker, J. M. Healey, and H. T.

Engelhardt, Jr., *The Law-Medicine Relation: A Philosophical Exploration*, D. Reidel Publishing, Co., Dordrecht, Holland, pp. 75–91.

[23]   Wade, O. L. and L. Beeley: 1976, *Adverse Reactions to Drugs*, 2nd ed., William Heinemann Medical Books, Ltd., London.

APPENDIX

B. INGEMAR B. LINDAHL

NOTES ON THE PHILOSOPHY OF MEDICINE IN SCANDINAVIA

1. INTRODUCTION

The philosophical foundations of modern medicine, both as science and clinical practice, has until recently attracted interest primarily in connection with ethical issues. One reason for this has been the technological advances of medicine, which constantly bring out new ethical and ethically related philosophical problems inherent in medicine. Medical decision-making, e.g., on abortion, euthanasia, transplantation surgery, in *vitro* fertilization, and artificial insemination, has been at the centre of attention. Since all ethical issues ultimately depend upon definitions of concepts, e.g., what is to be meant by 'meaningfulness', 'life', 'death', and 'natural rights' of human beings, conceptual analysis has been a natural component of the debate. During the past decade, however, an increasing number of philosophical studies have been devoted to different concepts, theories, and methodologies of medicine for more general scientific purposes, independent of particular ethical issues.

The present essay is an attempt to give a brief, general view of some recent developments in this latter field of philosophy of medicine in Scandinavia. This outline will not attend to medical ethics, but will be confined to studies of topics like 'decision theory', 'medical classification', 'causality', 'concept formation', and discussions of different movements within philosophy of medical science and research; no attempt at completeness is made on any of these subjects. I shall begin by mentioning a few works within medical theory in a general sense, more or less pertinent to philosophy of medicine, and then concentrate on analysis with more traditional roots in philosophy.

2. THEORY OF MEDICINE

2.1. On Decision Theory

Besides the ethical interest of medical decision-making, which has been mainly concerned with the question of what constitutes a right course of action, other inquiries of philosophical significance have been made into the

L. Nordenfelt and B.I.B. Lindahl (eds.), Health, Disease, and Causal Explanations in Medicine, 237–248.

rational basis of the decision process itself. Among the major works are Dr Henrik Wulff's *Rational Diagnosis and Treatment* [53], Dr Greger Lindberg's *Studies on Diagnostic Decision Making in Jaundice* [26] and Mr Reidar Lie's *Acceptance and Rejection of Hypothesis in Medical Science* [24].

Wulff discusses a wide range of problems involved in clinical reasoning and decision-making. He analyses the different steps in the diagnostic process and comments upon the characteristics and reliability of the information obtained. He deals with the concepts of 'disease', 'causality', 'criteria of definition', and 'classification of diseases'. Different approaches to diagnosis, such as the probabilistic or Baysian way of reaching a diagnosis and methods for evaluation of diagnostic tests, are discussed. The book ends with a discussion of controlled and uncontrolled clinical trials.

Lindberg's study is mainly empirical, and he concentrates on a specific clinical problem. The aim of the study is to find ways of improving the efficacy of diagnostic decision-making for the diagnosis of jaundice. He uses mathematical models derived from the theory of statistical decision-making to assess diagnostic methods and the utility of data obtained from a group of jaundice patients in two Stockholm hospitals. The definition of diagnosis, the concept of disease, and the components of the diagnostic process are briefly discussed, and a concise orientation in the field of medical decision theory is given by way of introduction. Lindberg's work was preceded and partly inspired by a project on decision theory, led by Professor Gunnar Biörck, under the title 'The Medical Decision Making Process' ([26], p. 9). Biörck's own contributions to the analysis and discussion of the theory of medical decision-making dates back to the early 1960s, and comprise essays on various strategies and techniques for elucidating the structure of the diagnostic reasoning process ([2], [3], [4], [5]).

Lie examines two areas in cardiology, one from basic research and one from clinical science. He discusses problems germane to the rational acceptance of hypotheses in these areas. He criticizes the use of decision theory in clinical medicine and proposes a method based on the objective interpretation of probability and interval estimators. The need to understand how knowledge from both clinical trials and basic science is intergrated, in order to reach decisions in clinical medicine, is stressed. The concept of disease is also discussed.

A philosophical essay by Dr Gunnar Jonson [21], should also be mentioned here. Jonson applies Karl Popper's notions of refutation to diagnosis seen as a hypothesis about probability. By using Bayes' theorem the author derives rules for selection of diagnostic tests, by which the efficiency of forming diagnoses can be improved.

2.2. On Alternative Methods of Treatment

The increased interest in herbal preparations, anthroposophical and homeopathic medicines, and other methods of treatment not accepted in established medicine has given rise to theoretical and empirical analyses of philosophical relevance. The research designs for investigating the efficacy of therapeutical methods and medicines in established and unconventional medicine is discussed by Professor Olof Lindahl and Mr Lars Lindwall in *Vetenskap och Beprövad Erfarenhet* [25]. The theoretical and historical background of the placebo concept is described, and "the great value of the placebo effect in medical treatment" is emphasized. An empirical study of unconventional medicine has been carried out by Dr Nils-Olof Jacobson in *Naturläkemedel och Okonventionella Behandlingsmetoder* [17]. The aim of his investigation is to describe the use of naturopathic medicines and to study the relationship between such use and the experiences of and attitudes towards conventional sick care. Patients and physicians in Sweden were studied. Jacobson discusses the difficulties of defining the basic concepts of his investigation, "naturopathic medicine", and "alternative methods of treatment". He points out the historical relativity of the latter concept; he describes the origin and development of currently "unconventional" methods of treatment, their organizations, legislation, and the fundamental features of the debate and extensive literature in this area of medicine.

2.3. Symposia

The Academy of Finland organized, in 1979, a seminar on the idea and image of man in medicine, *Ihmiskuva Lääketieteessä* [43], and a European symposium on medical theory, *Research on Health Research*, [14]. In the seminar the idea of man was discussed with reference to ethical aspects and their relevance for, e.g., medical education, research and the concept of disease. The image of man was also among the topics of the European symposium: methodological, ontological and conceptual aspects were discussed. A criticism of the Popper—Eccles' model of man, and a discussion of new possibilities to explain the relationship between man, his mind, and environment from a modern neuroscientific point of view were presented. The symposium contained several philosophical, sociobiological, ethnomedical, technological, educational, and health policy questions. Among the philosophical contributions were lectures on the problem of causality in biomedical research; models for describing and explaining the development of medical science and scientific knowledge; the ontological concept of disease; disease defined in terms of subnormal and subnatural organic function; and analogies between the concept of mental and somatic health.

3. PHILOSOPHY OF MEDICAL SCIENCE

3.1. Education and Textbooks

For decades, philosophy has had an unique position in the academic education in Norway, Denmark, and Iceland in the form of the so-called 'Philosophicum': A one term obligatory course in philosophy and behavioral science for all students entering the universities. This program is still on-going in Norway. It was abolished in Denmark in 1971 and Iceland in 1979, but this tradition had given philosophy a natural place in medical education. Philosophy of science and research is among the obligatory subjects in the current plan for medical education in Sweden, which was adopted in 1977 [46]. Philosophy is also taught at the medical schools in Finland. In all five countries, however, the teaching of the philosophy of medicine still only amounts to some ten or twenty hours in basic medical training. In recent years, on the research level, courses in philosophy of medical science have been held at various intervals in the larger university cities in all the Scandinavian countries.

The need for a body of literature in the teaching of philosophy of medical science has resulted in three text books on the subject: *Medicinsk Videnskabsteori* by Drs Anders Ottar Jensen and Hans Siggard Jensen [18], *Fra Filosofi til Fysiologi* by Professor Knut Erik Tranøy [45], and *Medicinsk Vetenskapsteori* by Dr Germund Hesslow [15]. They represent three different approaches to the subject. Jensen and Jensen concentrate on the different movements or traditions within philosophy of science (i.e., positivism, hermeneutics/phenomenology, and Marxism), and discuss some central ideas in medical science in relation to these traditions; Tranøy offers a historical approach and includes medical and odontological ethics; and Hesslow deals mainly with methodological issues, e.g., analysis of probability estimations, causality, hypothesis-testing and definitions.

Philosophy of medical science is also treated by Professor Uffe Juul Jensen in *Videnskabsteori 2* [19], and Professor Torgny Sjöstrand in *Medicinsk Vetenskap* [39]. In contrast to the positivistic tradition, Jensen stresses the social and historical context, and the dialectical interaction between theory and practice in which science develops. Sjöstrand surveys the history of ideas in medicine from antiquity to modern times, and advocates a controversial view that modern philosophy is distinct from science.

An introduction to medical concept formation and semantics is given by Dr Helge Malmgren in *Vetenskapsteoretiska Aspekter på Medicinsk Begreppsbildning* [27], and a collection of essays in philosophy of medicine has been compiled by Mr Nils Gilje, Mr Reidar Lie and Professor Gunnar Skirbekk [13].

3.2. On Analysis of Specific Issues

(a) *The concepts of health and disease* have, as fundamental notions in medicine — and for their decisive influence on the role and autonomy of man in society in general — been subject to a number of studies from different perspectives.

Dr Katie Eriksson analyses the health concept in *Hälsa* [11] from a pedagogical point of view. She discusses criteria for a descriptive definition of health based on theories and definitions of health in dictionaries and international nursing literature. Dr Nordenfelt approaches the concept of health (as formulated by Caroline Whitbeck [48], and Ingmar Pörn [36]) in *The Equilibrium Theory of Health* [33], by scrutinizing the notions of *goal* and *ability*, utilizing distinctions within a philosophical theory of action.

In the psychiatric debates concerning the concepts of 'mental health' and 'disease' specifically philosophical contributions appear not infrequently. Among the recent essays are a discussion of free will and mental health by Gunnar Skirbekk [41], and an analysis of the psychopathological disease concept by Alf Ross [38]. They both accomplish their analyses of the demarcation between mental health and disease in terms of "communication competence". Skirbekk supports a normative concept of health, focusing on linguistic communication competence as a central characteristic. Ross develops a normative concept of mental disease, defined by functional disorders in the psychophysical organism, which comprises the interpersonal (social) communication as a principal criterion. Ross has also pointed out the normative aspect of the somatic disease concept in an earlier essay [37].

Georg Henrik von Wright's axiological treatise *The Varieties of Goodness*

[49] contains a discussion of medical goodness and the related notions —
health and illness. A being is in bodily health, according to von Wright, when
its organs perform their proper function, and it suffers from illness when the
proper function of the organs (or an organ) is impeded and causes it *pain* or
pain-like sensations, such as discomfort, ache, nausea (directly or indirectly,
through causing *incapacitation* and frustration of wants). Mental health and
illness are similarily characterized by the function of faculties of the mind
(except for the pain or discomfort component).

The concepts pain and incapacitation are also the basis for Nordenfelt's
analysis of the disease concept in *Till Sjukdomsbegreppets Semantik* [30]. His
analysis shows that a disease concept defined solely by pain and incapacitation
becomes both too inclusive and too exclusive, as many diseases are not
associated with pain or incapacitation, and certain forms of pain or incapacitation are not connected with what we would like to call diseases. Nordenfelt
distinguishes between a "clinical concept" (illness), presupposing pain and
incapacitation, and a "theoretical concept" (disease) based on our intuitions
about biological functioning. A major source in Nordenfelt's analysis of this
distinction and in his discussion of the concepts of somatic and mental
disease is found in the work of Christopher Boorse ([6], [7], [8]).

Boorse's [8] value-neutral, biologically and statistically oriented concept
disease is compared to Georges Canguilhem's [9] evaluative concept (tied
to feelings of suffering and impotence) in Nordenfelt's *On the Pathology of
Georges Canguilhem* [32].

Another pair of disease notions are examined by Dr Wulff in his essay on
the essentialistic view of diseases as entities existing independently of the
sick patient [55]. Wulff criticizes the essentialistic (or "demonic") disease
concept and defends a nominalistic (or "patient-oriented") notion of disease.
(See also [53]).

Uffe Juul Jensen analyses in *Sygdomsbegreber i Praksis* [20] the philosophical and methodological positions advocated by the various groups in the
clinical collective (medical doctors, nurses, therapists, etc.). He argues that
no single method can solve the basic problems which the clinical collective
faces: (1) how to draw a line between cases that should be treated and cases
that should not be treated, and (2) how to choose between different kinds
of treatment. After having analysed the different *philosophical* positions,
Jensen turn to problems of *practice*. He conducts a kind of Wittgensteinian
analysis of the different kinds of clinical practice (e.g., the specialized clinic,
primary health care, community medicine, psychiatric clinic). He describes
and delimits the different kinds of procedures used in different practices

as a basis for critical evaluation and development of the various kinds of clinical practice.

The concept of disease has also been the subject of two trans-disciplinary symposia: the Swedish Society of Medical Sciences' symposium *Sjukdoms-begreppet – Behövs det?*, 1974 ([42], [44]), and a seminar at the Department of Social medicine, University of Copenhagen, *Sygdomsbegrebet – Praktiske, Politiske og Teoretiske Betragtninger*, 1978 [12]. Clinical, epistemological, sociological, and social insurance aspects were discussed at the Swedish symposium. The Danish seminar covered research – methodological, social political, ethnomedical, preventive medical, ontological, and epistemological aspects germane to the concept of disease.

(*b*) *Classification*. Closely associated with the questions of defining diseases in general and the demarcation of particular ailments are the problems of classification. The newly issued International Classification of Impairments, Disabilities, and Handicaps (ICIDH) [52] is critically examined, and constitutes the point of departure for the development of a conceptual framework for understanding disabilities and related concepts, in Nordenfelt's *On Disabilities and their Classification* [35]. The eighth revision of the Manual of the International Statistical Classification of Diseases, Injuries, and Causes of Death (ICD) [50] (which is still in use in the Nordic countries) is discussed both in *Om Grunden för Svensk Dödsorsaksstatistik* [31] by Nordenfelt and Ingemar Lindahl, and in *Causes of Death* [34] by Nordenfelt. The latter study also deals with the latest, ninth revision of ICD [51], and with early models of this classification. The central issue in these two studies ([31], [34]) is, however, not classification itself but the conceptual and causal basis for the international cause-of-death statistics.

(*c*) *Causality and Causal Explanations*. Nordenfelt and Lindahl's [31] account for the legislation and practical procedure for collecting mortality data, and they map out conceptual and causal theoretical problems inherent in the international instructions for establishing the primary or underlying cause of death, and for recording on the medical certificate form the principal course of events leading to death. These problems of causality are further discussed by Nordenfelt in [34], which also contains an analysis of probabilistic treatments of causation.

3.3. Philosophy of Psychiatry

(*a*) *Ideals of science and clinical methodology*. Two views dominate the animated discussion of theories and methodologies in psychiatry: On the one hand, a biologically, neurophysiologically oriented way of seeing cases of mental disease, characteristically combined with more somatic and pharmacological methods of treatment; on the other hand, a more psychological and psychotherapeutic way of approaching patients. These two views are expressed in the two traditional positions within the philosophy of science, i.e., positivism and hermeneutics.

A major issue in the current debate, extending over the past decade, is the legitimacy of *hermeneutics* as a viable theory of psychiatric investigation.

Briefly, hermeneutics, with its roots in the humanities, and with its emphasis on the investigator's interpretive understanding (as opposed to causal explanation of the phenomenon he or she investigates), is judged by the majority of its critics (as well as many of its proponents) as containing elements alien to, and even incompatible with natural science. Among the most distinct critics of hermeneutics are those who regard these departures from natural science as tantamount to departures from the ideals of science as such ([16], [40]). Yet several of the proponents of hermeneutics point out methodological similarities to natural science, e.g., the application of a hypothetical-deductive method, as the decisive scientific trait ([10], [22], [23]). An effort has also been made to formulate criteria of "valid" or "true" interpretation or understanding, analogous to the formation of criteria of *valid* or *true* knowledge in natural science [1].

Some basic issues within philosophy of science of special relevance to psychology and psychiatry are discussed by Dr Malmgren in *Förstå och Förklara* [28]. Among other topics, he analyses different ideals of explanations and the epistemological basis for clinical observations.

(*b*) *Psychoanalysis.* The development of the psychoanalytic discipline, its basic concepts and theories, is examined from the point of view of its historical and social background and determining factors by Mr Arild Utaker in *Psykoanalyse og Samfunn* [47]. Dr Lennart Nilsson analyses the basis for psychoanalytic explanations with a similar approach in *Förklaringar inom Psykoanalysen* [29]. Nilsson discusses the origin and character of two basic views of the human psyche and its products, a humanistic and a naturalistic understanding. He makes a survey of the psychoanalytic theories about the human psyche and analyses two models of explanation inherent in these theories.

4. CONCLUDING REMARKS

As this short presentation suggests, scholarly attention has been given to a wide variety of topics and themes within philosophy of medicine in recent years. Attracting the greatest interest are medical concepts and semantical issues, with particular emphasis on the problems ingredient in decision theory, theories of clinical investigation and scientific explanations. Since the essays in this volume are already summarized in the 'Introduction,' I have generally refrained from mentioning them here (with the exception of Professor Pörn's contribution [36]), although they are all, as well as the Symposium itself, most relevant to the theme of this essay.

I have, then, distinguished between theory of medicine and philosophy of medicine, in spite of the fact that a significant part of the latter may be viewed as a subdiscipline of theory of medicine, and despite the fact that the borderlines of the general domain of philosophy are, at times, vague and indiscernible. Some works could of course be classified differently, since such classifications are at times somewhat arbitrary. It should further be noted that my selection principle has been primarily *topic* oriented and governed by the criterion of accessibility rather than by some geographically equitable distributive principle. (Sweden is obviously over-represented.) However, it has been my intention to include the most central studies in the broad field of philosophy of medicine throughout the Scandinavian countries.

Needless to say, clinical judgements and actions as well as purely scientific research activity have always been guided by philosophical preferences and assumptions. This then makes it necessary to be attentive to our philosophical presuppositions and to try to clarify them critically. This will become even more relevant as the practice of medicine continues and as reflection on the essential concepts of medicine continues among scholars. Like other sciences and technological advances, the capacity of medicine to influence and intervene will surely increase in the future. At the same time, the amount and specificity of the knowledge and the theoretical conditions for rational decisions will be significantly more complex. Accordingly, the need for theoretical stringency and critical reflection must be even more accentuated. The continuous interchange between philosophy and medicine portends the promise of a valuable contribution to this development.

Huddinge University Hospital,
Sweden

BIBLIOGRAPHY

[1] Aggernaes, A.. 1973, 'Hermeneutisk Metode I Psykiatrien' (with a Summary in English), *Nordisk Psykiatrisk Tidskrift* 27, 126–138.

[2]   Biörck, G.: 1963, 'Några Synpunkter på Diagnostikens Fysiologi', *Svenska Läkartidningen* 60, 311.

[3]   Biörck, G.: 1973, 'Statistics and the Individual Patient', *Skandia International Symposia, Early Phases of Coronary Heart Disease*, Nordiska Bokhandelns Förlag, Stockholm, Sweden, pp. 15–35.

[4]   Biörck, G.: 1974, *Den Medicinska Beslutsprocessen* (with a Summary in English), Spri Report 6/74, Spri, Stockholm.

[5]   Biörck, G.: 1977, 'The Essence of the Clinician's Art', *Acta Med Scand* 201, 145–147.

[6]   Boorse, C.: 1975, 'On the Distinction Between Disease and Illness', *Philosophy and Public Affairs* 5, No. 1.

[7]   Boorse, C.: 1976, 'What a Theory of Mental Health Should Be', *Journal for the Theory of Social Behaviour* 6, No. 1.

[8]   Boorse, C.: 1977, 'Health as a Theoretical Concept', *Philosophy of Science*, 44.

[9]   Canguilhem, G.: 1972, *Le Normal et le Pathologique*, Presses Universitaires de France, Paris.

[10] Elster, J.: 1979, 'Metoder og Prinsipper i Hermeneutisk Forskningstradisjon', *Nordisk Psykiatrisk Tidskrift* 33, 135–149.

[11] Eriksson, K.: 1977, *Hälsa, En Teoretisk och Begreppsanalytisk Studie om Hälsan och dess Natur som Mål för Hälsoedukation*, Institutionen för Pedagogik, Helsingfors Universitet, Helsingfors, Finland.

[12] Gannik, D., Jespersen, M., Krasnik, A., Launsø, L., Saelan, H. (eds.): 1978, *Sygdomsbegrebet – Praktiske, Politiske og Teoretiske Betragtninger*, Report 13, Institut for Social Medicin, Københavns Universitet, Denmark.

[13] Gilje, N., Lie, R. and Skirbekk, G. (eds.): 1982, *Kompendium for Filosofi-delen i Atferdsfag for Det Medisinske Studium ved Universitetet i Bergen*, Universitetsforlaget, Bergen, Oslo, Tromsø, Norway.

[14] Heikkinen, E., Vuori, H., Laaksovirta, T., Rosenqvist, P.: 1980, *Research on Health Research*, Proceedings of the European Symposium organized under the auspices of the European Medical Research Councils, Publications of the Academy of Finland 11, Helsinki, Finland.

[15] Hesslow, G.: 1979, *Medicinsk Vetenskapsteori*, Studentlitteratur, Lund, Sweden.

[16] Hesslow, G.: 1982, 'Bör Psykiatrin vara Naturvetenskaplig?', *Läkartidningen* 79, 209–212.

[17] Jacobson, N.-O.: 1979, *Naturläkemedel och Okonventionella Behandlingsmetoder. En Socialpsykiatrisk Undersökning av Erfarenheter och Attityder hos Läkare och Allmänhet* (with a Summary in English), Department of Psychiatry, Huddinge University Hospital, Huddinge, Sweden.

[18] Jensen, A. O., Jensen, H.S.: 1976, *Medicinsk Videnskabsteori*, Christian Ejlers' Forlag, København, Denmark.

[19] Jensen, U. J.: 1973, *Videnskabsteori 2, Ideologi og Videnskab, Sjaelelivet som et Socialt Produkt, Sygdom og Ideologi*, Berlingske Forlag, København, Denmark.

[20] Jensen, U. J.: 1983, *Sygdomsbegreber i Praksis, Det Kliniske Arbejdes Filosofi og Videnskabsteori*, Munksgaard, København, Denmark.

[21] Jonson, N. E. G.: 1982, *An Everyday Philosophy of Diagnosis* (manuscript), Surgical Department, Central Hospital, Kristianstad, Sweden.

[22] Jørgensen, F.: 1974, 'Den Hermeneutiske ('Forstående') Metodes Anvendelse Indenfor Psykiatrien', *Nordisk Psykiatrisk Tidskrift* 28, 4–11.

[23] Kringlen, E.: 1979, 'Metoder og Prinsipper i Psykiatrisk Empirisk Forskning', *Nordisk Psykiatrisk Tidskrift* 33, 150–161.

[24] Lie, R. K.: 1982, *Acceptance and Rejection of Hypotheses in Medical Science*, Report No. 65, Department of Philosophy, University of Bergen, Norway.

[25] Lindahl, O. and Lindwall, L.: 1978, *Vetenskap och Beprövad Erfarenhet*, Natur och Kultur, Stockholm, Sweden.

[26] Lindberg, G.: 1982, *Studies on Diagnostic Decision Making in Jaundice*, Departments of Medicine, Karolinska Institutet at Huddinge University Hospital and Karolinska Hospital, Stockholm, Sweden.

[27] Malmgren, H.: 1975, *Vetenskapsteoretiska Aspekter på Medicinsk Begreppsbildning*, Avdelningen för Vetenskapsteori, Umeå Universitet, Umeå, Sweden.

[28] Malmgren, H.: 1978, *Förstå och Förklara*, Doxa, Lund, Sweden.

[29] Nilsson, L.: 1979, *Förklaringar inom Psykoanalysen*, Akademilitteratur, Stockholm, Sweden.

[30] Nordenfelt, L.: 1979, *Till Sjukdomsbegreppets Semantik, En begreppsanalytisk Studie*, Akademilitteratur, Stockholm, Sweden.

[31] Nordenfelt, L. and Lindahl, I.: 1979, *Om Grunden för Svensk Dödsorsaksstatistik, Reflektioner kring Grundbegrepp, Regler och Praxis*, Department of Social Medicine, Huddinge University Hospital, Huddinge, Sweden.

[32] Nordenfelt, L.: 1982, *On the Pathology of Georges Canguilhem*, Department of Philosophy, University of Stockholm, Sweden.

[33] Nordenfelt, L.: 1982, *The Equilibrium Theory of Health, Health as the Equilibrium Between Ability and Goal, An Analysis of the Theories of Caroline Whitbeck and Ingmar Pörn*, Department of Philosophy, University of Stockholm, Sweden.

[34] Nordenfelt, L.: 1983, *Causes of Death, A Philosophical Essay*, Report 83: 2, Swedish Council for Planning and Coordination of Research, Stockholm, Sweden.

[35] Nordenfelt, L.: 1983, *On Disabilities and Their Classification, A Study in the Theory of Action Inspired by the International Classification of Impairments, Disabilities, and Handicaps (ICIDH)*, Department of Health and Society, Linköping University, Sweden.

[36] Pörn, I.: 1984, 'An Equilibrium Model of Health', in this volume, pp. 3–9.

[37] Ross, A.: 1979, 'Sygdomsbegrebet', *Bibliotek for Laeger* 171, 111–129.

[38] Ross, A.: 1980, 'Det Psykopatologiske Sygdomsbegreb', *Bibliotek for Laeger* 172, 1–23.

[39] Sjöstrand, T.: 1979, *Medicinsk Vetenskap, Historik, Teori och Tillämpning*, Natur och Kultur, Stockholm, Sweden.

[40] Sjöstrand, T.: 1982, 'Medicinens Naturvetenskapliga Grundval', *Läkartidningen* 79, 207–209.

[41] Skirbekk, G.: 1982, 'Den Frie Vilje og Psykisk Helse', Innlegg på den Nordiske Psykiater-kongress, 16 June 1982, Bergen, Norway.

[42] *Socialmedicinsk tidskrift*.: 1977, 54, 198–228.

[43] Suomen Akatemia: 1980, *Ihmiskuva Lääketieteessä*, Suomen Akatemian Julkaisuja 8/1980, Helsinki, Finland.

[44] Svenska Läkaresällskapet: 1974, 'Sjukdomsbegreppet – Behövs det?', Symposium

No 45, *Läkaresällskapets Riksstämma, Sammanfattningar*, 27–30 November, 1974, Stockholm, Sweden, p. 34.

[45] Tranøy, K. E.: 1978, *Fra Filosofi til Fysiologi, Filosofi, Naturvitenskap og Biomedisinsk Etikk – Utvalgte Tekster*, Universitetsforlaget, Bergen-Oslo-Tromsø.

[46] Universitets- och Högskoleämbetet: 1978, Utbildningsplan för Läkarlinjen, Fastställd av Universitets- och Högskoleämbetet 1977–03–18', *Samhällsmedicinska Moment m m i Läkarutbildningen, UHÄ-rapport 1978: 23*, Bilaga 3, Stockholm, Sweden.

[47] Utaker, A.: 1979, *Psykoanalyse og Samfunn*, Filosofisk Institutt, Universitetet i Bergen Norway.

[48] Whitbeck, C.: 1981, 'A Theory of Health', *Concepts of Health and Disease*, in A. L. Caplan, H. T. Engelhardt and I. J. McCarteney, Addison Publishing Company, Reading, Massachusetts, U.S.A.

[49] von Wright, G. H.: 1963, *The Varieties of Goodness*, Routledge and Kegan Paul, London, U. K.

[50] World Health Organization: 1967, *Manual of the International Statistical Classification of Diseases, Injuries, and Causes of Death*, based on the recommendations of the Eight Revision Conference, Geneva, Switzerland, 1965.

[51] World Health Organization: 1979, *Manual of the International Statistical Classification of Diseases, Injuries, and Causes of Death*, based on the recommendations of the Ninth Revision Conference, Geneva, Switzerland, 1977.

[52] World Health Organization: 1980, *International Classification of Impairments, Disabilities, and Handicaps, A Manual of Classification Relating to the Consequences of Disease*, Geneva, Switzerland.

[53] Wulff, H. R.: 1976, *Rational Diagnosis and Treatment*, Blackwell Scientific Publications, Oxford, U. K.

[54] Wulff, H. R.: 1979, 'What Is Understood by a Disease Entity?', *Journal of the Royal College of Physicians of London* 13, No. 4.

[55] Wulff, H. R.: 1981, 'The Disease Concept', in *Methodische Aanpak van klinisch Denken en Handelen*, Proceedings of a Boerhaave Cursus, Medical Faculty, Leiden, Netherlands, pp. 3–20.

NOTES ON CONTRIBUTORS

Anders Ahlbom, Ph.D., is Assistant Professor at the Department of Social Medicine, Huddinge University Hospital, Huddinge, Sweden.

Erik Allander, Ph.D., M.D., is Professor, Department of Social Medicine, Huddinge University Hospital, Huddinge, Sweden.

H. Tristram Engelhardt, Jr., Ph.D., M.D., is Professor, Departments of Medicine and Community Medicine; Member, Center for Ethics, Medicine, and Public Issues, Baylor College of Medicine, Houston, Texas, U.S.A.

Anne M. Fagot, Ph.D., M.D., is Assistant Professor of Philosophy, The University of Paris XII, and C.N.R.S., Paris, France.

Ralph Gräsbeck, M.D., is Professor, Minerva Institute for Medical Research, Helsinki, Finland.

Germund Hesslow, Ph.D. B.M., Department of Physiology and Biophysics, and the Department of Philosophy, Lund University, Lund, Sweden.

Uffe Juul Jensen, Ph.D., is Professor, Department of Philosophy, Aarhus University, Aarhus, Denmark.

Øivind Larsen, M.D., is Assistant Professor, Department of Medical History, University of Oslo, Oslo, Norway.

B. Ingemar B. Lindahl, B.A., is a doctoral student, Department of Social Medicine, Huddinge University Hospital, Huddinge, and the Department of Philosophy, University of Stockholm, Stockholm, Sweden.

Helge Malmgren, Ph.D., M.D., is Assistant Professor, Department of Philosophy, University of Stockholm, Stockholm, Sweden.

Lennart Nordenfelt, Ph.D., is Assistant Professor, Department of Health and Society, Linköping University, Linköping, Sweden.

Staffan Norell, Dr. Med. Sc., M.D., is Assistant Professor, Department of Social Medicine, Huddinge University Hospital, Huddinge, Sweden.

Ingmar Pörn, Ph.D., is Professor, Department of Philosophy, University of Helsinki, Helsinki, Finland.

Kazem Sadegh-zadeh, M.D., is Professor at the Institute of Theory of History of Medicine, University of Münster, Federal Republic of Germany.

Stuart F. Spicker, Ph.D., is Professor, Department of Community Medicine and Health Care, School of Medicine, University of Connecticut Health Center, Farmington, Connecticut, and Program Associate, Technology

L. Nordenfelt und B.I.B. Lindahl (eds.), Health, Disease, and Causal Explanations in Medicine, 249–250.
© 1984 *by D. Reidel Publishing Company.*

Assessment and Risk Analysis Group, PRA/STIA, National Science Foundation, Washington, D.C., U.S.A.

Henrik R. Wulff, M.D., is Chief Physician, Department of Medicine, Herlev University Hospital, Herlev, Denmark.

INDEX

ability
  health defined as 17–19
abnormality
  as a criterion of causality 149
  as a criterion of disease 16–17
abortion xxiv, 102, 104, 106, 231n, 237
adverse drug reactions xxix, 226–227
aging process 50
Ahlbom, Anders xxi–xxii, *93–98*, 99–100
Allander, Erik xxix, *215–223*
American Psychiatric Association 33
Aristotle xviii, 25, 45, 64, 114, 201
artificial insemination 38n, 237
association
  confounded 95, 97
  statistical 97–98

Bardwick, P. A. 120, 121
Bayes' theorem xxii–xxiii, 48, 101, 104, 238
Bayesian methods xxii–xxiii, 100, 101–125
Bernard, Claude 50, 111
Bichat, Xavier 28
bioethics 229–231
biological species xix, 64, 70–71, 73n
Biörck, Gunnar 238
Boorse, Christopher xiv, 29–31, 44, 188, 242
Braithwaite, Richard Bevan 99
Braybrooke, David 19
Broad street pneumonia 82, 83
Broussais, François J. V. 27–28

Cadman, Ed 228
Canguilhem, Georges xiv, 16, 242
Carnap, Rudolf 99
case control studies 94, 96
Cassell, Eric J. 215, 217

causal
  analysis xxi–xxiii, xxvi, 93–98, 99–100, 101–125
  chains 130, 144, 158–161
  classification 191
  explanation xxvii–xxviii, 63, 179, 183, 201–209, 211–212, 216, 220, 243
  law 206
  multiplicity xxiv, 143–154, 169–173
  responsibility 35, 102–107
  selection xxiii–xxvii, 137–150, 169–173, 180–182, 184–193, 193n
  strength 191
causal association 93, 191
  criteria of xxii, 96–97
causality
  concept of 201
  law of xx
  probabilistic account of 101
  probabilistic view of 107
causation
  models of 129–135
  sufficient/component model of xxv, 132–133
  web of xxiv, 131–132
cause
  definition of xx–xxi, 101, 170–172, 230–231
causes
  direct 143
  general versus singular 207–208, 212
  genetic 183–193
  initiating 145
  more/most important 145, 183–193
  necessary 104–107, 134, 170–173
  of death (*see* death)
  of disease (*see* disease)
  precipitating 137

251

causes (continued)
   primary xxv, 146
   principal 137
   selection of   xxi, xxv, 102–107,
      137–150, 184–193
   sufficient 104–107, 132–134, 170
   versus background conditions xxiii–
      xxiv, 35–36, 180
   weighting of xxi, xxv, xxvii, 184–
      193, 193n
classification
   of diseases xviii–xx, xxv, 55–58, 75–
      86, 153–163, 165–167, 173–
      177, 179–182, 238, 243
   of impairments, disabilities and hand-
      icaps 20, 243
   of patients 175
   psychiatric xx, 77–86
clinical
   chemistry xvii, 47–58
   diagnosis 68
   laboratory xvii, 47–58
   medicine 34–37, 37n, 45, 129, 149,
      238
   practice xvii, 45, 70, 208
   problems xvii, 34–37, 38n, 44, 75, ·
      172, 179, 238
   standard cases 69
   test standards 71
Clouser, K. D. 33, 38n
cohort studies 94, 96
Collingwood, Robin George 147–149
condition
   necessary xix, 101, 104, 106, 107,
      130, 133, 170, 185, 201, 225
   precipitating 188
   primary 40, 145–147
   sine qua non 106
   sufficient xix, xxiii–xxiv, 101, 104,
      130, 170, 185, 225
   underlying 138
confounding xxii, 95, 97
control
   group 94, 99–100
   individuals 47
Cullen, William 28
Culver, C. M. 33, 38n

Darwin, Charles 64
Davidsson, Donald 15
death
   causes of xxv, 137–150, 161
   certification form 138, 143–144
   definition of 150
   underlying cause of xxv, 137–150,
      161, 243
decision
   making 36–37, 180, 184, 230, 237,
      238
   theory 36–37, 115, 237, 239
definition
   Aristotelian theory of xviii–xix
   essentialistic 63–69
   operational xx, 77–86
diagnosis
   etiological (see etiological)
   levels of 56
   positive 114–117
   probabilistic 101, 109, 121
diagnostic
   criteria xx, 68, 77–78, 80–81, 89,
      93, 95
   operational criteria 77, 83, 89
   test standards 69, 71
Diagnostic and Statistical Manual of
      Mental Disorders (DSM-III) 78, 85
disability
   and non-ability 18
   and survival role 20–22
   criteria of 18–22
disease
   and incapacitation 242
   and pain 242
   as a cause of illness xv–xvi, 6–8, 11–
      13
   as a descriptive term xiv
   as an evaluative term xiv
   causes of 169–175, 181–182
   definition of xiii–xviii, 6, 27–41,
      54–55, 153, 241–243
   entities 33, 57, 63–65, 71–72, 75–76,
      173–174
   kind 64, 69–72
   manifestation of 67–68
   names xx, 78

disease (continued)
  non-evaluative/evaluative dimension
    of 29-30, 33, 43-44, 242
  objective 54-55
  subjective 54-55
  two doctrines of xiv-xv
  value-free concept of xvi
diseases
  as units of classification xix, 64, 70-
    72
  as units of evolution xix, 64, 70-72
  classification of (*see* classification)
  definition of xix-xx, 63-69, 75-76,
    77-86, 89-90
  development of 156-161
  essentialistic view of xviii-xix, 64,
    173, 242
  etiological understanding of 27
  genetic xxvii, 38n, 183-193, 195-
    197
  nominalistic view of 27, 173, 242
  ontological understanding of 27-29,
    43, 240
  physiological understanding of 27-
    29
  psychiatric 83-86, 89-90, 220, 241,
    242, 244
  standard cases of 67-69
dose-response
  relationship xxii, 97-98
Durand-Fardel 114

Eberle, R. 204
Eccle, John C. 239
von Economo, C. 79-80, 84
effects
  primary xxix, 225-231
  side xxix, 225-231
  synergistic 134
encephalitis lethargica xx, 79-81, 84, 90
Engelhardt, H. Tristram, Jr. xvi-xvii,
    xxix, 16, *27-41*, 43-45, 75,
    179-182, 225-233
entia morborum 27
epidemiological studies 93-98
epidemiology xxiv, 93, 129, 132-134,
    230

Eriksson, Katie 241
essentialism xix, 63-65
ethics, medical 50, 225-231, 237
etiologic fraction 134-135
etiological diagnosis xxii, 101, 102,
    107-115, 157
etiology
  multifactorial 175
  specific 84-85
  unitary 84-85
evolutionary biology 70
explanation
  causal (*see* causal)
  deductive-nomological xxvii-xxviii,
    102, 201-205, 211-212
  Hempel-Oppenheim's criteria (H-O
    scheme) of xxviii, 201-205
  scientific 202-203, 211
  statistical 208

Faber, Knud 27
Fagot, Anne M. xxi-xxiv, *101-126*, 184
Feinstein, Alvan 26, 64, 70-72
Forrester, J. W. 8n
Frankfurt, H. G. 9n
Fraser, David W. 83

Gert, E. 33, 38n
Gilje, Nils 241
Glover, Jonathan 231n
goals
  profile of xvi, 4-9, 11-13
Goosens, William K. I. 16, 29
Gräsbeck, Ralph xvii, *47-60*
Grene, Marjorie 30-31

Hart, H. L. A. 35, 149-150, 180
health
  as ability 15-22
  conceptual circle of 8n, 15-22, 25-
    26
  definition of xii-xviii, 3-9, 11-13,
    15-23, 25-26, 48-49, 153,
    217, 241-243
  equilibrium model of xvi, 3-9, 11-
    13, 241
  longitudinal assessment of 50

health (continued)
  mental 240-242
  molar view of 11, 16-17, 25-26
  molecular view of 16-17, 25
  objective 49-50
  public (see public)
  subjective 49-50
  versus illness xv-xvi, 6-8, 11-13
Hempel, Carl G. xxvii-xxviii, 99, 201-
  207, 211
hermeneutics 244
Hesslow, Germund xxvii, *99-100, 183-
  193*, 195-197, 240
Hill, A. Bradford 96-97
Hippocrates 226-227, 231n
history
  of concepts of disease xxv-xxvi,
    27-30, 173-175
  of medicine 161, 162, 173-175
holistic
  approach 49, 216
  medicine xxix, 215-222
homosexuality
  as a disease 31-33
Honoré, A. M. 35, 149-150, 180

illness
  definition of xv-xvi, 6-8, 11-13
Imhof, Arthur E. 163n
Ingelfinger, Franz 229
International Classification of Impair-
    ments, Disabilities, and Hand-
    icaps (ICIDH) 20, 243
International Federation of Clinial
    Chemistry 47
International Form of Medical Certifi-
    cate of Cause of Death 138,
    143-144
INUS-condition (INUS-factor) 63, 170-
    177, 196-197

Jacobson, Nils-Olof 239
Jensen, Anders Ottar 240
Jensen, Hans Siggard 240
Jensen, Uffe Juul xix, *63-73,* 75-76,
    240-242
Jonson, Gunnar 238

Kass, L. 38n
Kaufman, A. S. 19
Kendell, R. E. 84
Kierkegaard, Søren Aabye 7
Koch, Richard 27
Koch, Robert 207
Korsakow's amnesia 85

laboratory
  medicine 47-58
  tests xvii, 48
lactose intolerance xxvii, 191
lacunae (of brain) 113-118
Laënnec, René Théophile Hyacinthe
    173
Lange, F. 121
Larsen, Øivind xxv-xxvi, *153-164,*
    165-167
Lasagna, Louis 229
law
  causal 206
  deterministic 206, 211-212
  of coexistence 203
  of succession xxviii, 203, 205-206
  statistical 208
Legionnaire's disease xx, 81-84, 90
Lie, Reidar K. 238, 241
Lindahl, B. Ingemar B. xxv, *137-152,
    165-167, 237-248,* 243
Lindahl, Olof 239
Lindberg, Greger 238
Lindwall, Lars 239
von Linneaus, Carl 28, 64, 173
Lunn, Villars 65, 67-68

MacKay, D. M. 8n
Mackie, John L. xxi, 63, 169-172, 174,
    181
Malmgren, Helge xx, *77-87,* 89-90, 241,
    244
manipulability
  as a causal criterion 147, 191-192
Manual of the International Classifica-
    tion of Diseases, Injuries, and
    Causes of Death (ICD) xxv,
    137-150, 151n, 243
Marxism 240

Maslow, Abraham H. 19, 22n
McCloskey, H. J. 19
medical
    certification form 138, 143–144
    education 240
    ethics 50, 225–231, 237
    history 161–162 (*see also* history)
    theory 237, 239
medical science 147, 240
    applied 179, 182
    applied versus unapplied 34–35, 37n,
        45
    clinical versus basic 34–35
    philosophy of 237, 240
medicine
    and law xxiv, 180
    history of (see history)
    philosophy of (*see* philosophy)
    preventive xxvi, 132, 155–163, 165,
        176–177
Meyrignac, C. 121
Mill, John Stuart xxi
Mill's method of induction xxi
modes of dying 149-150
modification rules 140–142
morbidity
    administered 156, 161, 165–167
    classified 155, 161, 165–167
    faces of 154–156, 165–167
    influenced 155, 165, 166
    perceived 155, 161, 163, 165, 166
    true 155, 165, 166
Morris, J. N. 148, 150n
mortality statistics 137–150, 161, 243
Murphy, Edmond A. 72, 193n

Nagel, Ernest 99
needs
    basic 8n, 13, 19–22
    Maslow's theory of 19, 22n
Nilsson, Lennart 244
Nordenfelt, Lennart *xiii–xxx*, xvi, 8n–
        9n, *11-13, 15-23,* 25–26, 151n,
        241–243
Nordic Clinical Chemistry Project
        (Nordkem) 58n
Norell, Staffan xxiv–xxv, *89-90, 129-135*

normal 187, 188, 193n, 195
    values 50
normality
    concept of 16, 30, 188, 193n

Occam's razor 205
Oppenheim, Paul xxvii–xxviii, 201–202,
        205, 211

Paracelsus 27
pattern recognition (causal) 102, 112–
        118
Peirce, Charles Sanders 179
Pharmaceutical Manufacturers Associa-
        tion 228
phenomenology 240
philosophy of medicine xiii, xvii, xx,
        184, 237–248
P.O.E.M.S. syndrome 120, 123, 124
Poirier, J. 114, 121
Popper, Karl R. 76, 99, 238, 239
prevention 129
    of death 137, 148, 153, 156–163
    primary 157
    secondary 148, 157
    tertiary 157
preventive medicine (*see* medicine)
probability
    a posteriori xxii, 101, 102, 110
    a priori xxii, 101, 110
profile of goals xvi, 4–9, 11–13
psychiatric classification (*see* classifica-
        tion)
psychiatry
    philosophy of 243–244
psychoses
    definition of 67–68
public health 137, 154–156, 161, 165,
        166, 215, 225
Pörn, Ingmar xvi, *3-9,* 11–13, *211-212,*
        241

reference
    individual xvii–xviii, 47–58
    values xvii, 47–58
Rhomberg, Ernst 37n

risk
  factors 101, 115, 137, 176
  relative 93
Ross, Alf 241
Russell, Bertrand xx

Sadegh-zadeh, Kazem xxi, xxviii, 34,
  43-45, 201-209, 211-212
Salmon, Wesley C. xxviii, 204-205, 208
Saris, N.-E. 47
de Sauvages, François Boissier 28, 32,
  173
Schaffner, Kenneth 180, 181
severity
  as a causal criterion 147
Shimpo, S. 119, 120
Sjöstrand, Torgny 241
Skirbekk, Gunnar 241
Spicker, Stuart F. xxix, 25-26, 225-
  233
statistical causal analysis xxii, 93-98,
  99-100
Stegmüller, Wolfgang 204, 208, 209n
Strömgren, E. 65
Suppes, Patrick xxii
Swedish Society of Medical Sciences
  243
Sydenham, Thomas 28, 64, 69, 72
syndrome 64-65, 69-72
synergy 130, 133-135

systems of actuality and of ideality
  3-7

Takatsuki, K. 120
Taylor, P. W. 8n
taxonomic species xix, 70-71
The Academy of Finland 239
The American Holistic Medical Associa-
  tion 217
The Scandinavian Society of Clinical
  Chemistry and Clinical Physiology
  54
Tranøy, Knut Erik 240

Utaker, Arild 244

Veatch, Robert 228
Virchow, Rudolf 28, 181

Wertheimer, R. 3
Whitbeck, Caroline 8n-9n, 11-13, 241
White, Morton 151n
World Health Organization xxv, 20, 48,
  54, 137, 138, 144, 145, 148-
  150, 217
von Wright, Georg Henrik 241
Wulff, Henrik R. xix, xxi, xxv-xxvi, 33,
  35, 65-66, 68, 71-72, 75-76,
  169-177, 179-182, 195-197, 238,
  242
Wunderlich, Carl 37n

The Philosophy and Medicine Book Series

Editors

H. Tristram Engelhardt, Jr. and Stuart F. Spicker

1. Evaluation and Explanation in the Biomedical Sciences
   1975, vi + 240 pp.   ISBN 90-277-0553-4

2. Philosophical Dimensions of the Neuro-Medical Sciences
   1976, vi + 274 pp.   ISBN 90-277-0672-7

3. Philosophical Medical Ethics: Its Nature and Significance
   1977, vi + 252 pp.   ISBN 90-277-0772-3

4. Mental Health: Philosophical Perspectives
   1978, xxii + 302 pp.   ISBN 90-277-0828-2

5. Mental Illness: Law and Public Policy
   1980, xvii + 254 pp.   ISBN 90-277-1057-0

6. Clinical Judgment. A Critical Appraisal
   1979, xxvi + 278 pp.   ISBN 90-277-0952-1

7. Organism, Medicine, and Metaphysics
   Essays in Honor of Hans Jonas on his 75th Birthday, May 10, 1978
   1978, xxvii + 330 pp.   ISBN 90-277-0823-1

8. Justice and Health Care
   1981, xiv + 238 pp.   ISBN 90-277-1207-7

9. The Law-Medicine Relation: A Philosophical Exploration
   1981, xxx + 292 pp.   ISBN 90-277-1217-4

10. New Knowledge in the Biomedical Sciences
    1982, xviii + 244 pp.   ISBN 90-277-1319-7

11. Beneficence and Health Care
    1982, xvi + 264 pp.   ISBN 90-277-1377-4

12. Responsibility in Health Care
    1982, xxiii + 285 pp.   ISBN 90-277-1417-7

13. Abortion and the Status of the Fetus
    1983, xxxii + 349 pp.   ISBN 90-277-1493-2

14. The Clinical Encounter
    1983, xvi + 309 pp.   ISBN 90-277-1593-9

15. Ethics and Mental Retardation
    1984, xvi + 254 pp.   ISBN 90-277-1630-7

16. Health, Disease, and Causal Explanations in Medicine
    1984, xxx + 250 pp.   ISBN 90-277-1660-9